Building Research Establishment Report

Energy use in buildings and carbon dioxide emissions

L D Shorrock, BA, PhD and G Henderson. BSc, MSc

Books are ⸱⸱⸱ ⸱⸱⸱ ⸱⸱ned on or befor⸱

Building Research Establishment
Garston
Watford
WD2 7JR

Price lists for all available
BRE publications can be
obtained from:
Publications Sales
Building Research Establishment
Garston, Watford, WD2 7JR
Tel: Garston (0923) 664444

BR 170
ISBN 0 85125 436 5

696
SHO

CONTENTS

SUMMARY

This report explores the relationship between energy use and the emission of the major greenhouse gas carbon dioxide. The emission, since 1950, of carbon dioxide due to the burning of fossil fuels is considered both globally and for the United Kingdom. This enables the more detailed analysis of the United Kingdom carbon dioxide emission considered here to be viewed with the global perspective in mind. It is shown that, whilst the global emission of carbon dioxide has risen dramatically, the emission by the United Kingdom has remained fairly static. It is concluded that about half of the total United Kingdom emission can be attributed to the energy used by buildings and that about 60% of this is attributable to dwellings.

The potential for reducing carbon dioxide emission through the application of energy efficiency measures in dwellings is investigated, assuming that current patterns of primary energy use for heat and power generation do not change. It is shown to be possible to reduce the carbon dioxide emission currently associated with dwellings by 35% using proven technologies. Furthermore, it is shown that the reduction due only to energy efficiency measures that are currently cost-effective is 25%. About two-thirds of these savings are due to improved insulation standards, with the remaining third being due to improved appliance efficiencies.

The broad conclusions obtained for the United Kingdom are drawn together as general statements which should be relevant to many industrialised nations. The methodology used to pinpoint key target areas for reducing United Kingdom carbon dioxide emissions could, in principle, be adapted to any country, using currently available knowledge.

INTRODUCTION

The 'greenhouse effect' is the name given to the phenomenon whereby trace gases in the atmosphere absorb infra-red radiation emitted by the Earth's surface, causing a warming of the atmosphere. It is this natural effect which is responsible for maintaining temperatures at the Earth's surface that allow life to flourish. In the absence of the effect, surface temperatures would be so low that the oceans would freeze and life as we know it would be impossible. The actions of mankind, however, in striving for ever-improving standards of living, have disrupted the natural balance by adding extra 'greenhouse gases' to the atmosphere. There is firm evidence[1] that concentrations of greenhouse gases are growing. In the case of carbon dioxide, the main concern of this report, atmospheric concentration has grown from about 280 parts per million (ppm) in the middle of the nineteenth century to about 350 ppm now. This 25% increase over about 100 years represents an average growth rate of about 0.2% per year. The current growth rate is about 0.5% per year. Taken with the growth of other greenhouse gases, current climate model predictions are broadly consistent with the observed global mean temperature rise of about 0.5°C over the past hundred years[1].

The main greenhouse gases being emitted are carbon dioxide (from burning fossil fuels), methane (largely from agricultural activities), chlorofluorocarbons (CFCs; used in aerosol sprays, refrigeration equipment and certain insulants) and nitrous oxide (mainly from internal combustion engines and nitrogen-based fertilisers). The CFCs, besides being extremely potent greenhouse gases, are also implicated as being responsible for the depletion of the Earth's protective ozone layer, and urgent action has already been taken to curtail emissions of CFCs. Of the remaining greenhouse gases, carbon dioxide is of special concern because its production is inexorably linked to maintaining and improving standards of living since it is directly related to the use of fossil-fuel-derived energy. Carbon dioxide is the largest single waste product of modern society and it is thought to account for about half of the warming effect associated with greenhouse-gas emissions. Later in this report the scope which exists within the United Kingdom (and, in particular, within the buildings sector) for reducing carbon dioxide emissions through the application of energy efficiency measures will be assessed.

This report concentrates on exploring the relationships between energy use and carbon dioxide emissions. Some additional statistics on other greenhouse gases, about which less is known at present, are provided in Appendix A. In Part 1 of the report the historical perspective is discussed, both for the whole world and for the United Kingdom. In Part 2 the focus is on the United Kingdom and on identifying the contributions of individual sectors. For the buildings sector, the analysis is taken further by considering the contributions of individual end uses of energy in dwellings. This analysis then allows some estimates to be presented for the potential reductions to carbon dioxide emissions through the application of energy efficiency measures in dwellings.

It is important to realise from the outset that this is a subject area where there are inherent uncertainties in the values that are calculated. The use of energy statistics to calculate carbon dioxide emissions requires a number of assumptions to be made, particularly with regard to calorific values and carbon contents of fuels. The assumptions used for this report are discussed in Appendix B. It is possible to make different, but equally plausible, assumptions, which will lead to slightly different results. The broad conclusions that are reached in this report, however, should remain unaffected by minor variations in the assumptions.

PART 1 HISTORICAL PERSPECTIVE FOR THE WORLD AND THE UNITED KINGDOM

World energy use and carbon dioxide emissions since 1950

Table 1, which is drawn from UN energy statistics[2,3], gives the primary energy use, by fuel, of the major world regions from 1950 to 1985. Figure 1 illustrates how the primary fuel mix has changed as world energy consumption has risen, whilst Figure 2 illustrates how each of seven major world regions have contributed to this growth. It should be noted that values which appear in Figures 1 to 6 for the years between 1950 and 1955, 1955 and 1960, 1960 and 1965, 1965 and 1970 were obtained by linear interpolation. The figures quoted for 1981 in Table 1 were also obtained by linear interpolation, because data for this year were not present in the statistics available to us. UN energy statistics data for years after 1985 were also not available when this report was being prepared. The figures were originally quoted in tonnes of coal equivalent and have been converted into petajoules (PJ) as explained in Appendix B. The figures for primary electricity are expressed, as is the normal practice in the United Kingdom, in terms of the notional fossil fuel input to equivalent contemporary conventional steam stations. A nominal thermal efficiency of 30% has been assumed, which is the value used in the UN energy statistics. This does not affect the calculated carbon dioxide emissions but does serve to indicate the significance of primary electricity in terms of fossil-fuel generation.

It is clear that primary energy consumption has risen in all world regions, but, in relative terms, the developing regions of the world (Asia in particular) have increased their primary energy use by much more than the developed regions have. In fact, in the most developed regions of the world, energy consumption by 1985 was either rising very slowly or had stabilised, this being a result, in part, of the response to the oil crisis in the 1970s. There are, of course, considerable variations in the rates of increase within the individual countries which make up the seven world regions considered. Use of gas, oil and primary electricity has grown at a much faster rate than that of solid fuel.

Table 2 gives the figures from Table 1 converted into carbon dioxide emissions. (See Appendix B for a discussion of the assumptions which were used to make this conversion.) Note that, for each unit of energy obtained, solid fuel produces about 75% more carbon dioxide than gas does, whilst liquid fuel produces about 50% more carbon dioxide than gas does. In contrast, primary electricity (geothermal, hydro, nuclear and wind) produces no carbon dioxide. There has been a growth of over 200% in global carbon dioxide emission due to fossil-fuel burning since 1950. Global carbon dioxide emission due to fossil-fuel burning has now reached about 20 billion tonnes per year. Non-fossil-fuel sources are not considered in this report, but it is worth noting that the use of traditional fuels (fuelwood, bagasse, etc) adds (assuming it is not replenished) approximately 2 billion tonnes to this total, whilst deforestation contributes a further 3 billion to 9 billion tonnes[1]. The largest part of the increase has occurred, as illustrated in Figure 3, as a result of the growth in the use of liquid fuel (oil). Figure 4 illustrates the same global growth in carbon dioxide emission sub-divided into the seven world regions. This figure shows some interesting trends. It is clear, for example, that Western Europe has dropped from second to fourth in the ranking of carbon dioxide emission by region, having been overtaken by Eastern Europe/USSR and Asia. Most significantly, the carbon dioxide emission of Asia has grown by almost 14 times and looks set to overtake Eastern Europe/USSR and USA/Canada before the end of the century.

Perhaps the most worrying aspect of this is that, dramatic though the growth in Asia's carbon dioxide emission has been, the emission per capita (see Figure 5) is still tiny in comparison with that of the more developed world regions (it is less than a tenth of the emission per capita in USA/Canada). It is worth noting also, that Eastern Europe/USSR has shown by far the greatest absolute increase in carbon dioxide emission per capita. Both Western Europe and USA/Canada appear to have passed through a peak in the carbon dioxide emission per capita which occurred in the mid- to late 1970s. The carbon dioxide emission per capita of USA/Canada, however, remains the highest, at over twice that of Western Europe and one and a half times that of Eastern Europe/USSR.

A useful measure of the amount by which changes to the primary fuel mix have contributed to changes to carbon dioxide emissions is provided by the carbon dioxide emission per primary PJ. These figures are illustrated in Figure 6, which shows that increased reliance on non-solid-fuel sources has meant that all world regions have improved by this measure. The region which has shown the least improvement is Asia. In contrast, there has been a very large improvement in Western Europe. A similar large improvement also shows clearly in the corresponding figures for the United Kingdom, and this is discussed in the next section.

United Kingdom energy use and carbon dioxide emissions since 1950

Table 3, which is drawn from the Digest of United Kingdom Energy Statistics[4,5], gives the United Kingdom consumptions of primary fuels for each year since 1950. (See Appendix B for a discussion of the assumptions which have been made to convert from the original figures, quoted in million tonnes of coal equivalent, into PJ.) As shown in Figure 7, the primary energy consumption of the United Kingdom rose steadily up to the early 1970s. During this period the use of coal declined, whilst the use of petroleum grew. The oil crisis in the early 1970s marked the beginning of changes to this pattern. Energy conservation became a priority, and this was accompanied by changes to the fuels being used. Subsequently, the use of petroleum declined again, the use of coal remained fairly constant, whilst the use of natural gas grew rapidly. Primary electricity use grew steadily throughout the whole of the period under consideration. Since the early 1970s the net result of all these changes has been a slight decline in total primary energy use (which would probably have been more pronounced but for some exceptionally cold winter weather during 1979, 1986 and 1987).

Figure 8 illustrates the effect of these changing fuel consumptions on the carbon dioxide emissions of the United Kingdom. These figures are also given in Table 4. (See Appendix B for a discussion of the assumptions used to convert the energy consumptions into carbon dioxide emissions.) The total carbon dioxide emissions are similar to those quoted in the Digest of Environmental Protection and Water Statistics[6]. There are some differences between the estimates, however, which should be noted. The first difference is that the figures in this report refer only to energy-related carbon dioxide emissions whilst the Digest figures include other sources (the Digest figures for total carbon dioxide emission are, therefore, marginally higher than the figures in this report). The second difference is a matter of definition. In this report, as explained in Part 2, the carbon dioxide emissions of power stations are allocated amongst the electricity consumers. In the Digest, the emissions from power stations are treated separately. It is for this reason that the figures for carbon dioxide emissions of buildings given in the Digest are much lower than the figures quoted in this report.

Although primary energy consumption rose quite rapidly up to the early 1970s, the rise in carbon dioxide emission was much less pronounced. Carbon dioxide emission peaked at about 700 million tonnes per year in the early 1970s. It has since fallen off by about 15%, currently having reached about 600 million tonnes per year. This is about the same as the level of emission in the early 1960s. Indeed, the most notable feature of the United Kingdom carbon dioxide emission has been its remarkable constancy. On a per capita basis, as shown in Figure 9, there has been a reduction in emissions since the broad peak of the 1970s, much as has occurred in USA/Canada and Western Europe (Figure 5). Carbon dioxide emission per capita is now almost the same as it was in the early 1950s.

It is clear that the changing primary fuel mix of the United Kingdom has been a major factor in holding carbon dioxide emissions at a relatively constant level. This is shown most clearly by the figures for carbon dioxide emission per primary PJ (Figure 10). This large improvement parallels that observed for Western Europe as a whole (Figure 6).

In Part 2 of this report the present carbon dioxide emission of the United Kingdom is examined in more detail, and this leads to conclusions about what can be done to reduce these emissions further. The United Kingdom presently only accounts for about 3% of the world carbon dioxide emission due to fossil-fuel burning. Although the greenhouse effect is a global problem, it is still instructive to consider what an individual country, like the United Kingdom, can do to reduce its carbon dioxide (and other) emissions. Many of the broad conclusions reached would probably be equally valid for other countries in the OECD (Organisation for Economic Co-operation and Development). Furthermore, the methodology used for the analysis is almost certainly adaptable to any country, assuming that the relevant data exist for that country.

PART 2 ANALYSIS OF PRESENT UNITED KINGDOM CARBON DIOXIDE EMISSION, AND POTENTIAL FOR ITS REDUCTION THROUGH ENERGY EFFICIENCY IN BUILDINGS

Introduction

In order to analyse what role energy efficiency can play in reducing the carbon dioxide emissions of the United Kingdom, it is important to focus on the end uses of energy (demand) rather than the supply. This means that, although the carbon dioxide emission associated with electricity generation actually occurs at fossil-fuel-fired power stations, it must be allocated amongst the electricity consumers. If this is not done, there is no way in which one can readily determine the carbon dioxide reductions that would result from energy efficiency measures implemented by the electricity consumers. Focusing on the end uses of energy, of course, means using <u>delivered</u> energy statistics rather than the primary energy statistics used in Part 1. In order to calculate carbon dioxide emissions from delivered energy, however, it is necessary to determine the primary energy equivalent of the energy delivered. This is particularly important for electricity, where the 'primary energy ratio' (defined as the ratio of the total primary energy input to the energy delivered to consumers) is necessarily large because of the inherently low thermal efficiency of conventional power stations. For the other delivered fuels it is possible to calculate carbon dioxide emissions directly from the calorific value and carbon content of the fuel. To do so, however, requires more assumptions to be made about calorific values and carbon contents of fuels. In this report, to be consistent with Part 1 and to minimise the number of assumptions, all fuels are related to their primary energy equivalents (coal, oil, gas or primary electricity) through their primary energy ratio and primary fuel mix. (See Appendix B for further discussion of this point.)

The end uses of energy can be described at many levels. Identifying the sectors to which energy is being supplied is the most aggregated level. Within each sector one can then attempt to identify individual end uses of energy (eg for buildings these could be space heating and cooling, water heating, cooking, lighting and appliances). If the data are available, it is even possible to go beyond this level to consider how the individual end-use energy consumptions are made up (eg appliances could be sub-divided into washing machines, tumble-driers, refrigerators, freezers, etc). Generally speaking, information is only available at the more aggregated levels and anything beyond this is more speculative. For dwellings, however, there is sufficient information available to be able to identify, with some confidence, exactly those end uses given as examples above. (Of course, in the United Kingdom there is effectively no cooling load in dwellings since the peak temperatures experienced are rarely high enough to cause serious discomfort.) This is possible because the results of research work and practical experience gained over many years have allowed models of energy use in individual dwellings and in the whole United Kingdom housing stock to be developed. The models developed by the Building Research Establishment (BRE) are called BREDEM[7] (BRE Domestic Energy Model - which considers individual dwellings) and BREHOMES[8] (BRE Housing Model for Energy Studies - which considers the whole housing stock). BREHOMES is a model of energy use in the housing stock which divides the stock into representative categories of dwelling (of which there are over 400) defined by built form, age, tenure, and ownership of central heating. BREHOMES uses the BREDEM model to calculate the energy use of each category of dwelling. Results from the BREHOMES model are used in some of the following sections of this report.

BRE's future programme includes the development, within the next few years, of an equivalent of BREHOMES for the non-domestic building stock. For this report, however, the emphasis is on the domestic sector (dwellings), although the energy use and associated carbon dioxide emissions of the whole building stock are also estimated. In addition, the overall energy use and carbon dioxide emissions of each of the other main sectors are calculated.

United Kingdom energy consumption and carbon dioxide emission in 1987

In 1987 the energy delivered to final consumers in the United Kingdom amounted to about 6190 PJ. This energy was delivered in a number of forms, as listed in the first column of Table 5. In order to deliver this quantity of energy to final consumers a greater amount of primary energy was consumed, totalling about 9110 PJ. This 9110 PJ was made up of the quantities of primary fuels (coal, oil, gas, nuclear electricity, hydro-electricity and imported electricity) shown at the bottom of Table 5, which are in close agreement with the 1987 totals quoted in Table 3. (The reason for the imperfect agreement is discussed in Appendix B. Essentially, the Table 3 figures are more approximate than those of Table 5.) The ratio of primary energy used to delivered energy is called the primary energy ratio, and it gives an indication of the amount of energy used in the processing and distribution of fuels. In 1987 the average primary energy ratio was 1.47 (Table 5).

It is clear from Table 5 that the primary energy ratio differs considerably between fuels. For coal, for example, delivery to final consumers involves very few losses, and so the primary energy ratio is close to unity. On the other hand, there are large losses associated with generating and distributing electricity, and so electricity has a high primary energy ratio. Partially compensating for these losses is the fact that electricity has a very high 'end-use efficiency', whilst solid, liquid and gas fuels have somewhat lower end-use efficiencies. In addition, of course, electricity is uniquely suited to certain end uses and so it cannot simply be displaced by other fuels with more favourable primary energy ratios. Given the amount of carbon dioxide produced by each of the primary fuels (see Appendix B) and the equivalent primary fuel mix of each of the delivered fuels (Table 5) it is possible to calculate the carbon dioxide emission per PJ associated with each of the delivered fuels. These figures are shown in Table 6. Obviously, electricity is the largest producer of carbon dioxide because of the high primary energy ratio and the fact that it is mostly generated by burning fossil fuels. Using the delivered energy figures in Table 5 together with the figures in Table 6, it is easy to show that electricity accounts for the largest share of the total United Kingdom carbon dioxide emission, closely followed by petroleum. The percentage shares of each of the main categories of delivered fuel are illustrated in Figure 11.

The Digest of United Kingdom Energy Statistics[4,5] provides information on the total use of fuels by various sectors. The main categories defined are industry (iron and steel), other industry, transport, domestic, public administration, agriculture and miscellaneous. The main interest in this report is to deduce the energy use, and consequent carbon dioxide emission, of buildings. It is actually rather difficult to draw the line between what constitutes building energy use and what does not. Nevertheless, it is clear that it must include nearly all the energy use of the domestic and public administration sectors (and probably most of the miscellaneous category) and some of industry's energy use, but very little of the energy use of the transport and agriculture sectors. With this in mind, the energy use of buildings has been defined as all the energy use of the domestic sector, all the energy use of the public administration and miscellaneous sectors, and a fraction of the energy use of industry, but none of the energy use of the transport or agriculture sectors.

It is not possible, using the information within the Digest of United Kingdom Energy Statistics, to determine the fraction of industry's energy use which can be considered to be building related. However, there is information available[9] which does allow the appropriate fraction to be estimated. The data used to make this estimate are somewhat out of date now (they refer to 1980) but, bearing in mind the broad approximations which have had to be made for the other sectors, they are sufficiently accurate for our purposes. In future, when a non-domestic equivalent to BREHOMES has been developed, it will be possible to improve the accuracy with which overall building energy use is estimated. The building-related energy use of the industry sector has been derived assuming that it is made up of energy used for space heating, water heating, lighting and appliances. It shows that about 20% of the total energy used by industry goes towards building-related uses. This percentage varies considerably between different fuels and between the two industry sectors defined in the Digest of United Kingdom Energy Statistics.

Energy consumption and carbon dioxide emission attributable to the United Kingdom building stock in 1987

Table 7 gives the delivered energy consumptions of the various sectors within the United Kingdom for 1987. Figure 12 illustrates the breakdown of energy consumption by sector, whilst Figure 13 illustrates the breakdown by sector and by fuel. It will be noted that buildings account for very nearly half of the energy delivered to all sectors of the United Kingdom. Furthermore, buildings use about two-thirds of the electricity consumed in the United Kingdom and, hence, they account for more than half the primary energy use of the United Kingdom. (Note that own generation of conventional thermal electricity by industry is not counted as electricity consumption – this appears in the statistics as consumption of those fuels used for generation.)

This reliance on electricity means that buildings are ultimately responsible for almost half of the United Kingdom carbon dioxide emission. Table 8 gives the breakdown of carbon dioxide emission by sector and by delivered fuel. This breakdown is illustrated in Figures 14 and 15 which are the carbon dioxide emission equivalents of Figures 12 and 13.

Tables 7 and 8 give, respectively, figures for the energy consumption and carbon dioxide emission of buildings. The buildings in those tables are divided into four categories. The domestic category (dwellings) is the most important in terms of both energy use and carbon dioxide emission. The domestic sector accounts for about 60% of both the energy use and the carbon dioxide emission attributable to buildings. Fortunately, the domestic sector is also the easiest building sector to analyse, owing to the existence of the BREHOMES model. Unlike dwellings, which conform to a relatively small number of built forms with fairly standard sizes and constructions, the non-domestic sector varies widely in character. Shops, for example, vary between extremes such as the small 'corner shop' and the huge department store. Offices, factories and other non-domestic buildings cover similarly wide ranges. Not surprisingly, this makes non-domestic buildings much more difficult to model than domestic buildings. Such modelling is, consequently, still in its infancy and a 'broad brush' approach is the best that can be done in the near future. For the non-domestic sector it is assumed, for the time being, that overall percentage savings could be comparable to those obtained for the domestic sector.

Energy consumption and carbon dioxide emission attributable to United Kingdom dwellings in 1987

Table 7 gives the total consumption of each fuel by the domestic building sector (dwellings). In order to determine how much of each of these fuels goes towards particular end uses, it is necessary to refer to the results obtained from the BREHOMES model. Table 9 gives the BREHOMES estimated breakdown of the end uses of each fuel, which are also illustrated in Figure 16. The values quoted are actually based on 1986 results scaled to the 1987 total consumptions. Note that the different types of solid fuel have, for convenience, been amalgamated into one category in the figure and the table.

From the values in Table 9 and Figure 16, it will be clear that space heating accounts for most of the energy used by dwellings and that, for this end use, natural gas is the dominant fuel. A similar pattern applies to the next largest end use, water heating. This is not surprising in view of the fact that space and water heating demands are quite often met by the same heating appliance. Electricity, however, has a greater share in meeting the water heating demand than it does for space heating. After space and water heating demands, the next largest energy demand is that for lights and appliances (note that 'appliances', as defined here, does not include cookers), which, for all practical purposes, can be considered to be met entirely by electricity (non-electrical household appliances such as gas refrigerators do exist but only in small numbers). This end use accounts for slightly more than half of the total electricity use of dwellings. The cooking load (not counting appliances such as microwave ovens, toasters, etc, which are included within the lights and appliances total) is largely shared between electricity and gas.

Table 10 and Figure 17 give the carbon dioxide equivalents of Table 9 and Figure 16. The importance of the electricity used for lights and appliances becomes quite apparent. It contributes almost the same carbon dioxide emission as does all the gas burnt for space heating purposes. It is worth pointing out that the fuel mix for generating electricity has been assumed to be the same for all end uses. In reality, there would be a higher nuclear component at night, and so end uses which rely on off-peak electricity (such as space heating with night storage heaters) would contribute somewhat less carbon dioxide per unit of electricity consumed (calculations suggest about 15% less with the current fuel mix) than end uses which rely more on daytime electricity (such as lights and appliances). Individual appliances, depending on time of use, may contribute either more or less than the average carbon dioxide emission per unit of electricity consumed, but overall it is reasonable to assume that the average figure applies. Any error, due to this assumption, in the electricity-related carbon dioxide emissions of the end uses shown on Figure 17 must be below 15%, and is likely to be much less than this.

Obviously, the energy consumption of dwellings for lights and appliances must be a priority target for energy efficiency improvements. This is particularly so because lights and appliances consumption is rising at a rate of about 5% per year as households acquire a wider range of appliances. There are, in fact, significant energy savings, and attendant reductions of carbon dioxide emission, which can be made for all the end uses identified in Tables 9 and 10 and Figures 16 and 17. In the following section some estimates are presented of the possible energy savings and reductions to carbon dioxide emission which can be achieved through the application of several energy efficiency measures.

Possible energy savings and reductions to carbon dioxide emission for existing United Kingdom dwellings

Energy-saving measures in dwellings

Energy savings in dwellings can be made in many ways. Space heating consumption can be reduced by insulation of the external fabric of a dwelling (including draught proofing) and/or by installing heating appliances with a higher efficiency. Water-heating energy savings can be made by insulating hot water storage tanks and, as with space heating, by installing higher efficiency heating appliances. For lights, appliances and cookers, energy savings can be made by choosing products with higher efficiencies. There are, in addition, a number of other means of saving energy which can be grouped together under the general heading of 'good housekeeping'. These include actions such as choosing appliances which are suited to the household composition (for example, choosing a freezer which is the right size for the household and which can be kept well stocked), turning off lights in unoccupied rooms, keeping doors between rooms closed, etc. These 'good housekeeping' measures are difficult to quantify, since they depend entirely on the behaviour of the household's occupants. Nevertheless, it makes good sense to promote public awareness of these simple measures which help to avoid wasteful use of energy. In what follows, only the reasonably quantifiable of the most common energy efficiency measures will be considered. More advanced energy efficiency measures which are presently less common, but which may well have an important role to play in the future, particularly in new dwellings, are not included. Examples of such measures include mechanical ventilation heat recovery, better utilisation of solar energy, advanced lighting and heating controls, etc. The better utilisation of solar energy, however, merits at least a brief discussion since it is the subject of intensive research effort in the United Kingdom and elsewhere.

Better utilisation of solar energy

The attraction of solar energy is, of course, that it is a clean, virtually inexhaustible, source of energy. Buildings can take advantage of solar energy in a number of ways. Essentially, however, there are two broad categories of solar technology which are relevant to buildings — active and passive techniques.

Active solar techniques involve the use of collectors to capture energy from the sun and, usually, some means of transferring the energy to a store for later use. Such systems are generally used for water heating and space heating and cooling applications. It has been shown that active solar techniques are not, at present, economically attractive in the United Kingdom[10]. Generally speaking, it is only in situations where large quantities of tepid water are required (eg swimming pools) that the economics of active solar techniques begin to look attractive, at current price relativities. The other active technique of importance is the use of photo-voltaic collectors which can generate electricity directly from the sun's energy. Again, at present, these are not generally cost-effective for the buildings sector.

Passive solar techniques rely on using a building's form and fabric to capture, store and distribute solar energy. Nearly all buildings benefit from passive solar gains, by virtue of the fact that they have windows. The existing United Kingdom housing stock receives, over a heating season, about 330 PJ of solar energy, which, together with metabolic gains, offsets some of the delivered energy requirement by reducing the need for space heating. Figure 18 shows how these natural gains contribute to the overall delivered energy requirement of the housing stock. Solar energy is seen to form about 14 to 15% of the total delivered energy. The total delivered fuel, of course, is the same as shown in Figure 16 and Table 9. It is worth noting that the widespread uptake of double glazing would slightly reduce the passive solar contribution in the existing housing stock.

The aim of passive solar design is to optimise the contribution of passive solar gains. The orientation and distribution of glazing is, therefore, a particularly important aspect of passive solar design. As such, it is particularly relevant to the design of new buildings, although there are some passive solar components (for example, conservatories) which can be retro-fitted. Estimates suggest that the national primary energy savings which could be achieved by passive solar design of dwellings would, after 20 years of gradual introduction of the technology, amount to between about 5 and 10 PJ per year[11]. In terms of carbon dioxide emissions this corresponds to a reduction of between 0.4 and 0.8 million tonnes per year. It will be noted that this is rather small compared with the savings that are discussed in the following sections of this report. However, it must be remembered that the immediate national benefit of any energy efficiency improvements which are largely restricted to new buildings (for example, the introduction and implementation of new Building Regulations) will be relatively small in the short term. This is because new building forms only a small part of the existing stock - the new-build rate is about 1% of the existing stock per year.

Cost-effectiveness of energy efficiency measures

The cost-effectiveness of any energy efficiency measure is an important consideration. In this report, therefore, a distinction is drawn between those measures which are presently cost-effective and those which are not (but which are still technically possible). There are several ways of measuring the cost-effectiveness of a particular energy efficiency measure and many different criteria can be defined to determine what is cost-effective and what is not. It is beyond the scope of this report to discuss the measurement of cost-effectiveness or to define cost-effectiveness criteria rigorously. The aim, rather, is to give an indication of the difference in achievable savings between what are generally accepted to be cost-effective energy efficiency measures and those measures which are technically possible. The report by Pezzey[12] gives a thorough discussion of the cost-effectiveness of different energy efficiency measures. For the purposes of this report, the following broad definitions have been adopted:

1 The term **'technically possible'** is taken to mean what could actually be done now if cost were no obstacle. It does not, however, include retro-fitting of measures which would normally be so disruptive that they would not generally be considered (eg floor insulation). In the case of household appliances such as cookers, refrigerators, etc, it should be noted that 'technically possible' has been moderated somewhat by the view that what is technically possible is unlikely actually to happen (at least within a relatively short period of time) unless there is market intervention in the form of, for example, appliance labelling schemes. Thus, whilst it is probably feasible[13] to improve the efficiencies of major appliances by as much

as 50%, a 20 to 25% improvement is more likely. For refrigeration equipment it is also probable that the technically feasible improvements could, ironically, be unattainable because of environmental concerns restricting the use of certain CFCs. The alternative refrigerants currently being considered are about 10% less efficient than those which have traditionally been used. Similarly, the external skin of refrigerators and freezers may well have less effective insulation properties when alternatives to CFCs are used. These are further reasons for assuming rather conservative technically possible improvements for household appliances.

2   The term 'cost-effective' is taken to mean what is technically possible and can be shown to be a good investment. For insulation measures there is some difficulty in deciding where one draws the line between cost-effective and not cost-effective. For example, it is well established that adding insulation to a loft which has no existing insulation is a good investment, with a payback time of, typically, under 2 years. It is much more difficult, however, to decide whether it is cost-effective to 'top up' existing loft insulation since this will depend, amongst other things, on the thickness of the existing insulation, on current (and future) fuel prices and on the criteria adopted for cost-effectiveness (eg the maximum acceptable payback time). At present, it is generally accepted that it is worthwhile topping up loft insulation if it is only 30 mm or less thick. If the existing insulation is 50 mm thick, however, the decision is not clear-cut. For a measure such as double glazing there are even greater difficulties in assessing the cost-effectiveness. In purely energy-saving terms double glazing is rarely cost-effective but it has other benefits which make it attractive to householders, and it is generally viewed as a good investment because it undoubtedly adds to the market value, or at least the ease of selling, of a dwelling. For this reason there is justification for classing double glazing as a cost-effective energy efficiency measure. In addition, the marginal cost of double glazing above single glazing is such that, when windows have deteriorated to the extent that they have to be replaced, double glazing then is often cost-effective in energy-saving terms alone.

For household appliances it is often the case that what is technically possible is also cost-effective. For example, there are refrigerators on the market now which cost no more than other refrigerators but which are more efficient than their competitors. This can only be the case if the overcosts associated with manufacturing the more efficient refrigerators are small - implying that the efficiency improvement is both technically possible and cost-effective. For some other types of household appliance the cost-effective savings which could be achieved might be expected to be less than the technically possible savings. This may be because there is some barrier to their uptake (eg for low-energy light bulbs the high initial capital cost and the unusual physical appearance probably act as barriers) or because there may be a considerable lag in their uptake as a result of the relatively long lifetime of that type of appliance (so that old ones will continue to be used for some time to come). In these cases it has been assumed that the achievable cost-effective savings will amount to about half of the technically possible savings.

It will be clear from the above that the definitions adopted for what is cost-effective are rather broad, particularly for household appliances. It should also be borne in mind, however, that for household appliances the technically possible savings assumed are rather conservative and so the corresponding cost-effective savings are unlikely to be overestimates. Indeed, the assumptions made about appliances will, if anything, understate the importance of energy-efficient appliances in reducing carbon dioxide emissions. For the other energy efficiency measures the definitions of what is cost-effective are the result of discussions with the Energy Efficiency Office. Appendix B lists the assumptions used in this report concerning the quantification of technically possible and cost-effective savings.

Savings through insulation and draught proofing

There are about 21 million dwellings in the United Kingdom. The bulk of these were constructed before thermal insulation requirements were introduced into the Building Regulations. Nevertheless, in part as a result of Government initiatives and publicity campaigns, most homes now have much improved insulation levels[8,14]. The most notable success has been in the insulation of loft spaces. In 1974 about 40% of homes which could have loft insulation actually did have it. By 1987 this had grown to about 90% of homes. The thickness of installed insulation also improved - in 1974 the average was about 50 mm and in 1987 about 90 mm. The other most notable success has been in the insulation of hot water storage tanks - well over 90% of these are now insulated. Other insulation measures have also been taken up but to a lesser extent. By the end of 1987, 18% of cavity walls were insulated, 40% of homes had double glazing (though only 14% of homes had full double glazing), and about 40% of homes had draught proofing. It is clear, therefore, that there still exists a considerable potential for further improving the insulation standards of United Kingdom dwellings.

Table 11 shows the technically possible potential for energy savings in the present housing stock if the common insulation measures (including draught proofing) were to be taken up by all households which could benefit from them. It also gives the equivalent figures for currently cost-effective energy savings. It should be stressed that energy savings through insulation of a dwelling have, in the past, been difficult to predict for individual cases because occupants can choose to take a part of the possible energy savings in improved comfort standards. There are three aspects to comfort standards. One is the temperature level at which living-rooms are held during the evening. The others are the number of rooms that are heated on a regular basis and the time period for which they are heated. Changes in all three are governed by consumer expectations, by the availability of central heating and by economic considerations, including household income and cost of fuel. Between 1950 and 1980, 24-hour average temperatures in living-rooms increased at a rate of about 1°C per decade whilst evening living-room temperatures showed evidence for having reached a plateau by 1980[15].

As comfort standards improve it is more likely that the full potential energy savings will be realised. When comfort standards reach a ceiling (ie as indicated earlier, people have a limit on how warm they want to keep their homes) insulation will then achieve the full potential energy savings. The housing stock is now much closer to that ceiling than it was 10 or 15 years ago. This is reflected in the fact that about 70% of households now have central heating (about twice the percentage that had central heating in the early 1970s) and are, in general, attaining a good standard of comfort. In fact, this is one of the great successes of energy efficiency because, despite the tremendous improvements in comfort standards accompanying the use of central heating, there has been only a small increase in energy consumption by the housing stock since 1970, even though the number of households has increased by about 3 million. On a per household basis there has been little change in energy consumption since 1970, but average internal temperatures are estimated[14] to have risen by almost 3°C.

Table 11 shows that by far the biggest overall potential saving through insulation is that due to wall insulation. Note that in section (a) of Table 11 savings from insulation of all wall types are considered, whereas in (b) only those cavity walls which would be suitable for cavity wall insulation (assumed to be about 80% of cavity walls) are considered*. This is because (b) refers to cost-effective savings and, in most situations, solid wall insulation is not currently cost-effective. It should be noted, however, that solid wall insulation can be cost-effective when it accompanies major refurbishment work which has to be carried out anyway.

The overall savings for the other insulation measures are much lower, either because there is little remaining potential (as in the case of loft insulation and hot water tank insulation) or because the measure itself usually saves much less energy (as in the case of double glazing). The savings quoted for draught proofing are rather conservative for two reasons – first because it is a measure which is unlikely to be necessary where good quality replacement windows are installed (ie the windows incorporate draught seals), and secondly because, however good the draught proofing, there is a need to provide sufficient ventilation to avoid condensation and mould problems and to avoid the potential dangers where heating appliances require an air supply from within the dwelling.

Table 12 shows the reductions in carbon dioxide emission which would accompany the energy savings shown in Table 11 (assuming no change in the fuel mix). Insulation measures in the housing stock are seen to be capable of reducing the carbon dioxide emission of the United Kingdom by as much as 43.5 million tonnes, or about 7% of the present United Kingdom total emission. Cost-effective insulation measures could reduce emissions by almost 29 million tonnes, or about 5% of the present United Kingdom total emission.

Savings due to improvements to heating appliance efficiencies

In parallel with the rapid growth in the ownership of central heating there has been an improvement in average space heating efficiencies. This has occurred for a number of reasons. First, modern heating appliances are, in general, more efficient than their predecessors. Secondly, there has been a shift in the preferred fuel away from solid fuel and towards gas (meaning that open fires with typical efficiencies of only 30% have largely been replaced by much more efficient gas heaters and boilers). Thirdly, central heating is usually more efficient than individual heating appliances. Table 9 and Figure 16 show that heating by gas is dominant in the housing stock. Modern conventional gas central heating boilers are probably close to the maximum possible efficiency for such appliances. However, recent developments have led to gas boilers with even higher efficiencies coming onto the market. These boilers differ from ordinary boilers in that, rather than being designed to avoid condensation of flue gases, they are designed actively to encourage it, the resulting liquid usually being disposed of through a pipe to the dwelling's drainage system. Field trials[16] have shown that these condensing boilers can achieve seasonal efficiencies in the range 80 to 90% (compared with 65 to 70% for a typical conventional boiler). For this report it has been assumed, fairly conservatively, that the seasonal efficiency of a condensing boiler is 80%.

---

* Cavity wall insulation should not be installed where there is a high risk of water penetration from driving rain. Guidance on identifying suitable cases is given in the BRE Report 'Thermal insulation: avoiding risks', BR143.

By the use of a condensing boiler, the space and water heating consumption of a dwelling will obviously be reduced compared with its consumption where a conventional boiler is used. Table 13 gives estimates of the possible savings if all dwellings using gas central heating were to have a condensing boiler installed in place of their existing boiler. Note that the savings quoted assume that the insulation measures discussed above have also been carried out – if this were not the case the savings would be higher. This illustrates one of the characteristics of energy efficiency measures – namely that the savings for a particular measure within a series of measures depends on the order in which the improvements are carried out. If current standards of insulation are assumed, for example, the potential annual savings through the installation of condensing boilers <u>alone</u> amount to about 155 PJ (in terms of carbon dioxide, this represents a reduction of 8.5 million tonnes per year). Clearly, the energy savings associated with boiler replacement reduce with higher fabric insulation standards. It is for this reason that Table 13 shows that savings from boiler replacement are greater with 'cost-effective' insulation measures than with the more comprehensive 'technically possible' measures. In reality, of course, insulation improvements and boiler replacement often occur concurrently, or in reverse order, and it is not really possible to define uniquely the savings due to each measure separately – but the overall savings from the package of measures can be determined.

The fairly conservative savings quoted suggest that condensing boilers in housing might reduce the emission of carbon dioxide by as much as 5.5 million tonnes, or about 1% of the current United Kingdom total emission. Savings due to a general improvement in other heating appliance efficiencies are also likely to occur, but these will almost certainly be small in relation to the improvements which condensing boilers are capable of achieving. These other savings will not be considered here.

Savings due to improvements to lights and appliance efficiencies

Table 10 and Figure 17 show clearly that the electricity used for lights and appliances within the housing stock is a major source of carbon dioxide emission. Table 9 and Figure 16 give the corresponding total electricity consumption for lights and appliances. It is possible to estimate how much of this consumption is attributable to individual household appliances, as shown in Table 14, which is based on data from the electricity supply industry. It will be seen that there are certain major appliances that are responsible for a large proportion of the total consumption. In particular, appliances for refrigeration and washing or drying purposes consume a lot of electricity. Lighting, too, accounts for a large part of the total. The low ownership level of dishwashers means that, at present, these account for only a small part of the total consumption. In terms of consumption per appliance, however, dishwashers are heavy electricity users, and trends in other European countries and North America point to a strong growth in ownership.

Table 14 indicates the estimated technically possible improvement factors for the efficiencies of the major appliances relative to today's efficiencies, and gives the equivalent estimated cost-effective improvement factors. As noted earlier, the improvement factors assumed are deliberately rather conservative. They suggest that technically possible efficiency improvements to lights and appliances could reduce carbon dioxide emissions by about 11 million tonnes, or about 2% of the current United Kingdom emission. Cost-effective reductions are estimated to be about 8 million tonnes, or about 1.5% of the current United

Kingdom emission. A large part of these reductions is due to the increased use of low-energy compact fluorescent lights, which consume only a fifth of the electricity of conventional tungsten bulbs (hence the technically possible improvement factor of 75% given in Table 14). These lights are already widely available, although their present cost is such that they are usually only worth buying for rooms which are in fairly continuous use (ie living-rooms). In addition, the unusual appearance of the present generation of these lights probably acts as a barrier to their widespread uptake. There is no technical reason, however, why such lights should not be used in any room.

Savings due to improvements to cooker efficiencies

The efficiencies of cookers have improved over the years, due to such innovations as better controls (for example, spark ignition instead of pilot lights on gas cookers) and timers, faster-responding hot plates, better seals on doors, dual ovens, dual rings and grills, etc. Still further savings can be made by better insulation on ovens. In the 1970s it was estimated[13] that cookers, in line with other major appliances, could be improved in efficiency by about 50%. Figures from the electricity supply industry indicate that electric cookers have improved by about 10% over the years. Gas cookers have probably also improved by a similar amount. Therefore, a remaining technically possible improvement of 25% for gas and electric cookers, as indicated in Table 15, seems reasonable. For cooking by other fuels it has been assumed that there would be no improvement. The improvement factor for the equivalent cost-effective improvement figures (Table 15) is taken to be about half that assumed to be technically possible.

Table 15 shows a technically possible carbon dioxide reduction of about 3 million tonnes, or about 0.5% of the United Kingdom emission, and a cost-effective carbon dioxide reduction of about 1.5 million tonnes. These figures do not take account of the fact that cooking habits are changing. It is probable that convenience foods which use little energy in cooking will continue to be used more widely. It is also likely that microwave ovens will continue to grow in popularity. Greater use of microwave ovens would result in considerable reductions to the amount of energy used for cooking.

Summary of possible reductions to the carbon dioxide emissions associated with dwellings

Drawing together all of the estimated potential reductions to the carbon dioxide emissions associated with dwellings (Table 16 and Figures 19 and 20), it can be seen that very substantial reductions are feasible through the application of energy efficiency measures alone. All of the calculations have been done assuming that the consumption of the housing stock would otherwise remain static at its present level. This is obviously not entirely realistic (indeed the 5% per year growth of lights and appliances consumption mentioned earlier bears witness to this fact) but it does provide a firm baseline from which to work. It should be emphasised, therefore, that the carbon dioxide reductions quoted in this report are not forecasts of reductions in 'real time'. Future work will develop a number of scenarios to take the study beyond this baseline case which provides a first-order estimate of what carbon dioxide reductions can be achieved.

The overall technically possible carbon dioxide reductions amount to about 62 million tonnes (Table 16(a) and Figure 19). This is 35% of the total housing stock carbon dioxide emission or 10% of the total United

Kingdom emission. Over two-thirds of this reduction is attributable to cost-effective improvements. The overall cost-effective carbon dioxide reductions amount to about 44 million tonnes (Table 16(b) and Figure 20), or 25% of the total housing stock carbon dioxide emission and 7% of the total United Kingdom emission. If all savings that are not due to insulation are classified as being improvements to appliances (heating and cooking appliances are counted along with all other appliances) then it is interesting to note that improvements to appliances account, in both the technically possible and cost-effective cases, for about a third of the overall reductions to carbon dioxide emissions.

Other ways of generating heat and power

The savings calculated above are based on the assumption that the current patterns of primary energy use for heat and power generation do not change. This reflects the fact that the emphasis of the report has been on energy demand rather than energy supply. Changes to energy-supply practices, however, can have a substantial impact on carbon dioxide emissions (see, for example, Figures 7 and 8). For the purposes of providing the wider perspective, therefore, there follows a brief discussion of some supply options. It is important to note that calculated savings associated with alternative supply options cannot simply be added to the savings quoted in this report. There is an interaction between supply and demand which requires that, if both are to be considered, they must be considered together and not separately.

There are really two categories of supply options. One is the simple substitution of one delivered fuel for another - such as choosing to heat with gas rather than with solid fuel. This is a type of supply option that can readily be taken into account when creating scenarios such as those mentioned above, because recent trends in fuel use allow reasonable predictions of future preferences to be made. The other type of supply option is more difficult to deal with because it concerns, amongst other things, the choices made in the commissioning of power stations and the resulting changes to the efficiency of, and the fuel mix for, generation of electricity. It is the latter type of supply option that is considered in this section.

Table 5 shows that, in the United Kingdom, electricity is largely generated from coal. If more electricity were to be generated from oil or natural gas rather than from coal, then electricity would account for less carbon dioxide emission than it presently does. Primary electricity (nuclear, hydro, geothermal and wind power) produces no carbon dioxide and an increase in the use of such sources would also help to reduce carbon dioxide emissions. Of course, these sources are not without their own environmental and economic characteristics and so the scope for their wider introduction might be limited.

It is possible to increase the efficiency with which fuel is used to generate heat and power by using combined heat and power (CHP) systems instead of separate boilers and electricity generation plant. This is a technology for which the normally clear distinctions between supply and demand become blurred; indeed, because CHP can be used to supply energy requirements in buildings it can be considered to be an improvement in building energy efficiency. The overall efficiency of fuel utilisation with CHP can be as high as 85% - well above the 35% typical of conventional coal-fired power stations. Because of its superior fuel efficiency, CHP generation can make a significant contribution to reducing carbon dioxide emissions. However, its exact contribution will depend on a range of factors including the amount and type of plant installed, the fuel used, and the alternative energy-supply options displaced. Successive studies indicate considerable potential for the wider use of CHP technology to meet energy demands in buildings. This can be achieved either by using small-scale CHP systems on site or by connecting buildings to CHP/district heating schemes.

Another supply option technology which is particularly promising is that of combined cycle power plants. In such a plant, gas is burnt in a turbine which turns an alternator to generate electricity. Heat is also recovered from the gas turbine exhaust and is used to raise steam to drive a steam turbine. The steam turbine turns another alternator to generate more electricity. Overall thermal efficiencies above 45% are possible with this technology - considerably more than the 35% typical of conventional coal-fired power stations. A further advantage of this technology is its flexibility due to the fact that it is inherently modular.

Various technical options exist for cleaner and more efficient use of coal for generating heat and power. One such technology is fluidised-bed combustion which improves the efficiency with which coal is burnt whilst also reducing emissions of sulphur dioxide and nitrogen oxides. Such 'clean coal' technologies can be used in a combined heat and power mode. Coal can also be turned into gas to be used in a combined cycle plant which operates as described above, and such a plant is called an integrated gasification combined cycle plant. There are, in fact, several possible plant configurations which would result in cleaner and more efficient use of coal. It is possible, for example, to use fluidised-bed combustion in a combined cycle plant which operates in combined heat and power mode.

Discussion and conclusions

Figure 21 provides an overall summary of the impact that energy efficiency improvements in housing could have in reducing carbon dioxide emissions. A 25 to 35% reduction to the carbon dioxide emission associated with the housing stock is indicated, which is sufficient in itself to reduce the overall United Kingdom carbon dioxide emission by 7 to 10%. If similar savings can be achieved in other types of buildings, then a reduction of 12 to 17% to the United Kingdom carbon dioxide emission would be feasible solely through the application of energy efficiency measures in buildings. Application of combined heat and power technology could further add to this reduction. The quoted figure is based on the assumption that the energy consumption of buildings would otherwise remain static at its present level. As noted before, therefore, the figure is not a forecast of carbon dioxide emission reductions in 'real time'. It is, rather, a first-order estimate of what emission reduction could be achieved. The size of this estimate clearly indicates that substantial reductions to the overall carbon dioxide emission of the United Kingdom are possible through energy efficiency alone.

Gradual changes to the primary fuel mix would also help to reduce carbon dioxide emissions. As demonstrated in Part 1 of this report, changes to the primary fuel mix up to the early 1970s succeeded in holding United Kingdom carbon dioxide emissions fairly steady despite a 50% rise in energy consumption. Following the early 1970s, when the United Kingdom energy consumption peaked, continuing changes to the primary fuel mix helped to supplement energy efficiency improvements, resulting in a 15% reduction in carbon dioxide emissions. It is clear, therefore, that major improvements have occurred in the past, and this report has also demonstrated that there is ample scope for significant improvements in the future.

The conclusion from the above, therefore, is that, if the demand for useful energy were to remain constant, a target carbon dioxide reduction of 20% should be readily attainable through energy efficiency together with changes to the primary fuel mix.

There is no reason to suppose that the United Kingdom is very different from many other industrialised nations. Although there will be detailed differences (in the primary fuel mix, in the consumptions by particular sectors and in the consumptions for particular end uses) there will also be many similarities (the same primary fuels being used, the same broad sectors and the same end uses of energy). Indeed, the parallel between the improved carbon dioxide emission per PJ in the United Kingdom and in Western Europe as a whole (Figures 10 and 6) underlines the similarity of the United Kingdom to the rest of Western Europe. Similarly, Figure 5 suggests that Western Europe and USA/Canada have much in common, despite their very different levels of energy consumption per capita. These are just two examples which demonstrate that the broad conclusions obtained for the United Kingdom are probably equally applicable to many other industrialised countries. The broad conclusions concerning buildings can be summarised by the following general statements:

1 Buildings account for a major part of the carbon dioxide emission of an industrialised country.

2 Electricity, if mainly generated from fossil-fuel sources, will be a major source of carbon dioxide emission in an industrialised country. Buildings account for a large part of electricity consumption.

3 Reductions to carbon dioxide emission through the application of energy efficiency measures in buildings can be substantial.

4 Improvements to the efficiencies of appliances used in buildings will probably be almost as important as fabric insulation in contributing to reduced carbon dioxide emissions.

Irrespective of whether the above conclusions are generally applicable or not, it is certain that the methodology employed in this report to pinpoint key target areas and to quantify possible reductions to carbon dioxide emissions, could be adapted to any country. As noted previously, however, it is the availability of relevant data which determines how finely the end uses of energy can be identified. What this report has done is to demonstrate that the use of energy models (BREDEM and BREHOMES) allows the end uses of energy in dwellings to be determined with sufficient confidence that reliable estimates of potential reductions to carbon dioxide emissions can be made. Armed with this information it is then possible to evaluate different scenarios for the future and to develop appropriate strategies for action.

Related work

The work described in this report represents the first phase of the study. Related studies include:

1 **A non-domestic building sector model** to improve understanding of the scope existing within such buildings for improved energy efficiency and reduced carbon dioxide emissions.

2 **Time-based scenarios** to assess the impact of the rate of take up of energy efficiency initiatives on the reductions to carbon dioxide emissions.

18

3   **The contributions of other greenhouse gases** and their building-related use/emission will be better quantified. It is likely that the contributions of the other greenhouse gases will be described in terms of 'carbon dioxide equivalents'.

4   **Impacts on building design, fabric deterioration and energy use.** Up to this point the main concern has been the impact of buildings on the environment, by way of greenhouse-gas emissions. It must be recognised, however, that the environment also affects buildings, so that there are a number of 'building impact' problems associated with changes to the environment. The following examples indicate the kind of 'building impact' problems which will also be addressed:

    -   Climate change and climatic variability would be expected to change the requirements of building design, resulting in modifications to codes and standards, including Building Regulations.

    -   Some climatic changes could contribute to the deterioration of the fabric of buildings. For example, increased rainfall might result in increased risk of rain penetration and other moisture-related problems.

    -   A warmer climate would reduce the need for space heating in buildings, but it would also increase the need for air conditioning. It is not yet known (for the United Kingdom or, indeed, globally) whether the likely climate changes would result in a positive or negative feedback effect on energy requirements and on the emission of greenhouse gases.

APPENDIX A   OTHER GREENHOUSE GASES

In this appendix some extra information is given for other greenhouse gases. The information presented is far from exhaustive, reflecting the fact that much information has yet to be developed concerning these gases and their contributions to the greenhouse effect. At present, very little is known about the part which buildings play in contributing to the emission of these gases. It is certain, however, that buildings do play a part and, in what follows, the reader's attention will be drawn to the links between these gases and buildings.

As noted in the main text, carbon dioxide is thought to account for about half of the warming effect associated with greenhouse-gas emissions. The following table, which draws on a number of sources, gives estimated relative contributions of each greenhouse gas to the overall warming, together with other relevant information for each gas:

Greenhouse gas	Contribution to warming	CO_2 equiv/ molecule	Current conc	Growth rate (%/year)	Atmospheric lifetime
Carbon dioxide	50%	1	350 ppm	0.5	7 years
Methane	19%	30	1.7 ppm	1.0	10 years
CFC 12	10%	10 000	0.32 ppb	5.0	139 years
Tropospheric ozone*	8%	2 000	~20 ppb	0.5	Several weeks
CFC 11	5%	3 900	0.20 ppb	5.0	77 years
Nitrous oxide	4%	150	310 ppb	0.25	120 years
Water vapour	2%	–	–	–	–
Other CFCs	2%	–	–	–	–

* The lifetime of ozone in the troposphere is short but it is continuously being renewed and is growing in concentration. The growth rate is highly variable – 0.5% per year is fairly typical of reported figures.

Methane

The concentration of methane in the atmosphere has grown by a factor of roughly 2 in the past 100 years, and now stands at about 1.7 parts per million (ppm). This is a historical growth rate which greatly exceeds that of carbon dioxide. The current growth rate is about 1% per year (about twice that for carbon dioxide). Molecule for molecule, methane is about 30 times more effective than carbon dioxide as a greenhouse gas. Methane has a relatively long atmospheric lifetime of about 10 years.

Methane originates from a number of sources. Sources which have nothing to do with buildings include bacterial processes associated with ruminants, microbial activity in paddy fields and decay or burning of vegetation associated with agricultural activities. On the other hand, leakages of natural gas from pipelines are clearly related to buildings because most of the natural gas used goes towards building uses (see Figure 13). Similarly, methane emissions associated with coal-mining activities must also be related to the solid-fuel demand and, indirectly, to the electricity demand of buildings.

Chlorofluorocarbons (CFCs)

CFCs do not occur naturally but they are manufactured on a large scale.
The development of the two most common compounds, CFC 11 and CFC 12,
began in the 1930s. By 1985 the atmospheric concentration of CFC 11 had
reached 0.2 parts per billion (ppb), whilst CFC 12 had reached 0.32 ppb,
both with an annual growth rate of about 5%. Molecule for molecule, they
are several thousand times more effective than carbon dioxide as
greenhouse gases (CFC 11 - 3900 times, CFC 12 - 10 000 times). They also
have long atmospheric lifetimes (CFC 11 - 77 years, CFC 12 - 139 years).
Moreover, these gases are implicated as being responsible for depletion
of the ozone layer. Such is the concern about these gases that urgent
international action has already been taken to curtail emissions of
CFCs. Less harmful alternatives are currently under development.

The world demand (excluding USSR, Eastern Europe, and China) for CFC 11
and 12 in 1985 was about 750 000 tonnes[17]. Actual emissions of CFCs are
difficult to quantify because much of the use of CFCs involves a
'banking' of the gases in products which have a relatively long life (eg
freezers and refrigerators). Whether the gases are eventually released
to the atmosphere obviously depends on the arrangements made for
disposal. The primary applications of CFCs are in aerosols, foamed
plastics, refrigerants and solvents. There are two main applications
which involve buildings: foamed plastics used for insulation purposes in
buildings, and refrigeration and air conditioning. A further building-
related use is in certain aerosol foam sealants. In addition, halons,
which are used in the fire extinguishers present in many buildings, also
have similar properties to CFCs. (In fact, halons have an 'ozone
depletion potential' between 3 and 10 times that of the major CFCs.
Their 'greenhouse potential', however, is very uncertain.)

The following information regarding the use of CFCs in United Kingdom
buildings is known. Little is known about the resulting emissions of
CFCs. It should be borne in mind that this information is likely to
change as CFC alternatives become more widely used.

- About 80% of the insulation used in buildings is mineral-fibre based
  whilst about 9% is CFC-blown foam insulation[17]. This CFC-blown
  insulation includes polyurethane, polyisocyanurate, phenolic and
  extruded polystyrene foams.

- About 3000 tonnes of CFCs are used each year for manufacturing foam
  insulation for building purposes.

- Manufacture of domestic refrigerators and freezers uses 1400 tonnes
  of CFC 11 per year for the polyurethane foam insulation in their
  external skin. It is estimated that a maximum of 15 000 tonnes of
  CFCs are banked in the insulation of existing domestic refrigerators
  and freezers. A further 4500 tonnes are banked as refrigerants within
  existing domestic refrigerators and freezers.

- Manufacture of factory-insulated domestic hot water tanks uses
  approximately 110 tonnes of CFCs per year.

Nitrous oxide

The atmospheric concentration of nitrous oxide has grown from about
280 ppb in the nineteenth century to about 310 ppb in 1985. It is
currently increasing in concentration at a rate of about 0.25% per year.
Molecule for molecule, it is about 150 times more effective than carbon

dioxide as a greenhouse gas. It has a long atmospheric lifetime of about 120 years.

Nitrous oxide emission is related to the burning of fossil fuels and also to the use of nitrogen-based fertilisers. The principal fossil-fuel source is internal combustion engines. Building-related sources include domestic boilers and the electricity generation system.

Tropospheric ozone

The concentration of tropospheric ozone is very variable. Not only does it vary with latitude, altitude and location, but also with time of day. Typically, it is around 20 ppb, and is growing at about 0.5% per year. Molecule for molecule, it is about 2000 times more effective than carbon dioxide as a greenhouse gas. The excess production of ozone occurs by the photochemical action of sunlight on pollutants created by the burning of fossil fuels. Of particular importance are the pollutants from vehicle exhausts but the burning of fossil fuels to meet the energy demands of buildings also contributes.

APPENDIX B   ASSUMPTIONS MADE IN THIS REPORT

In this appendix the assumptions which have been made within the main report are explained in order to assist the reader in interpreting the results. Note that, because of the different statistical sources used, the assumptions for the world calculations are not exactly the same as those used for the United Kingdom.

Units

Energy can be measured in many different units, a selection of which appear in this appendix. This plethora of units inevitably contributes to much confusion and misinterpretation. Indeed, some of the units commonly used (for example, tonnes of coal equivalent) are almost useless for quantitative work unless they are supplemented by additional information on assumed calorific values (often not supplied), because they are masses of fuel rather than true energy units. For quantitative work it is preferable to use the internationally recognised Système International (SI) units. In the SI system the basic unit of energy is the joule (denoted by the symbol J). To indicate multiplication of a basic unit by a factor, a prefix is added. Multiplication by 10^{15} is indicated by the prefix peta (symbol P), leading to the unit used in this report – petajoules (symbol PJ). In Table 1 and Table 3 a related unit is also used – GJ (gigajoules = 10^9 joules). In this appendix, in connection with calorific values, the unit MJ (megajoules = 10^6 joules) also appears.

Other units used in the report should be familiar to all readers. It is perhaps worth noting, however, that the unit of mass used is the metric tonne. The difference between metric tonnes and tons is a further reason why units such as million tonnes of coal equivalent (symbol mtce) and million tons of coal equivalent (symbol mtce!) have not been used.

(continued)

Assumptions used for the world statistics

World regions

Seven world regions are discussed in Part 1 of the report. The regions chosen were largely dictated by the need to generate a continuous time series back to 1950. The regions used were based on the definitions in the 1980 UN Energy Statistics Yearbook[2] to which the reader is referred for details. Some of the regions have been amalgamated and some names altered to make them more readily recognisable, as follows:

Name of region	1980 UN Energy Statistics Yearbook equivalent
Africa	Africa developed + Africa developing
USA/Canada	North America developed
Rest of America	Other America developing
Asia	Far East developed + Far East developing + Middle East developed + Middle East developing + centrally planned Asia
Eastern Europe/USSR	Centrally planned Europe
Western Europe	Western Europe
Oceania	Oceania

Calorific values and conversion into PJ

The data used to draw up Table 1 were taken from tables in the UN statistics where the consumptions were expressed as tonnes of coal equivalent. The UN statistics quotes a 'coal equivalency' of 7000 calories per gram, which, in metric units, converts to a calorific value of 29.3076 MJ/kg. This calorific value was assumed to convert from tonnes of coal equivalent into PJ, and was also used in calculating carbon dioxide emissions. In the case of primary electricity, the conversion into PJ also included a division by a 30% nominal thermal efficiency (this is the efficiency which is quoted in the UN statistics) to obtain the notional fossil-fuel input which would be required to generate the electricity using conventional power stations.

For oil, the UN statistics give a figure of 1 tonne being equivalent to 10^7 kilocalories, which, in metric units, converts to a calorific value of 41.868 MJ/kg. This calorific value was used in calculating carbon dioxide emissions.

For gas, the UN statistics quote a 'standard heat value' of 39 021 kilojoules per cubic metre (39.021 MJ/m^3). If it is assumed that natural gas is almost entirely methane with a density of 0.717 kg/m^3, this becomes, on a weight basis, a calorific value of 54.4 MJ/kg. This calorific value was used in calculating carbon dioxide emissions.

Carbon contents of and carbon dioxide emissions from fossil fuels

The calorific value for coal given above is typical of bituminous coal. Bituminous coal typically has a carbon content of around 60 to 80%. Accordingly, it has been assumed that a carbon content of 70% is appropriate.

Oil is a complex mix of hydrocarbons. The carbon content varies according to the hydrocarbons present, but it is typically 87%, which is the value assumed in this report.

The composition of natural gas also varies. Usually, it is predominantly methane with relatively small amounts of other alkanes and trace quantities of other impurities. If natural gas were pure methane it would, from simple chemistry considerations, have a carbon content of 75%. The small presence of the other alkanes and impurities does not substantially change the overall carbon content, and so 75% has been assumed for this report.

Each carbon atom within a fossil fuel will produce one carbon dioxide molecule when the fuel is burnt (assuming complete combustion). Thus, knowing the carbon content of the fuel and the calorific value it is a simple matter to calculate how much carbon dioxide is produced for each unit of energy obtained by burning the fuel. Every 12 kg of carbon burnt produces 44 kg of carbon dioxide. Thus the following carbon dioxide emissions can be calculated:

Solid fuel: $(0.70 \times 44)/(12 \times 29.3076)$ = 0.0876 kg/MJ
                  = 0.0876 million tonnes/PJ

Liquid fuel: $(0.87 \times 44)/(12 \times 41.868)$ = 0.0761 kg/MJ
                  = 0.0761 million tonnes/PJ

Gaseous fuel: $(0.75 \times 44)/(12 \times 54.4)$ = 0.0505 kg/MJ
                  = 0.0505 million tonnes/PJ

These are the carbon dioxide emission factors assumed for the world statistics. Note that they differ slightly from the values assumed for the United Kingdom (see below). The important point to note is that solid fuel produces about 75% more carbon dioxide than gas does, whilst liquid fuel produces about 50% more carbon dioxide than gas does. The above figures for the carbon dioxide emissions of fossil fuels are very similar to values that have been quoted by other authors.

Note that these factors are applied to <u>primary energy</u> statistics. It is worth noting that calculation of carbon dioxide emissions from primary energy would be expected to produce an estimate which <u>may</u> be slightly on the high side because the losses between source and consumer do not necessarily involve burning the fuel. For example, some of the losses in delivery of natural gas are due to leakages from pipelines – which, in fact, would have a greater environmental impact than if the gas had been burnt because methane is a more effective greenhouse gas than carbon dioxide is! In reality, the uncertainties in the energy statistics themselves and in the assumptions about calorific values and carbon contents, mean that this is a source of error of only secondary importance. Furthermore, using primary energy statistics actually reduces the number of assumptions concerning calorific values and carbon contents. Moreover, international comparisons, of the sort in Part 1 of this report, are only readily done using primary energy statistics.

Assumptions used for the United Kingdom statistics

Delivered/primary energy considerations

Although the order in which the text of this report is presented is such that the history of United Kingdom energy consumption and carbon dioxide emission is presented before the detailed analysis of the 'present' situation, in fact the calculations were done the other way round. The reason for this is that the prime concern of this study has been to assess the scope for reducing carbon dioxide emissions from their present levels – only later did it become of interest to examine the historical perspective. This means that the calculations for the United Kingdom necessarily began from delivered energy statistics, in order that end uses of energy could be identified. All delivered energy consumption was related to its primary energy equivalent through the primary energy ratio and primary fuel mix.

Table 5 is a simplified energy balance for the United Kingdom, which allows one to deduce the primary energy equivalent of a given amount of a particular delivered fuel. It is based on the detailed energy balance for 1987 which appears as Table 6 in the 1988 Digest of United Kingdom Energy Statistics[5]. The total consumptions of primary fuels in Table 5 match those in the detailed energy balance (one small exception being that the oil figure has been adjusted to discount non-energy uses). For the time series from 1950, Table 1 of the Digest of United Kingdom Energy Statistics was used. This gives primary energy consumptions expressed in million tonnes of coal equivalent (using standard approximate conversion factors). In order to match the approximate figures for 1987 with the more accurate figures it is necessary to assume slightly different conversion factors between therms and tonnes of coal equivalent for the different fuels. The calculated conversion factors have then been applied to the figures for each year in Table 1 of the Digest of United Kingdom Energy Statistics to produce Table 3 of this report. The figures for primary electricity, as is the normal practice in the United Kingdom, incorporate a thermal efficiency factor so that primary electricity is represented by a notional fossil-fuel input to equivalent conventional power stations. There is a remaining slight discrepancy between the 1987 figures in Table 3 and Table 5 which is due to rounding of the figures (the calculated conversion factors were rounded, as were the original approximate primary energy consumptions).

Calorific values and conversion into PJ

To derive, from the Digest of United Kingdom Energy Statistics, appropriate calorific values for coal and oil, the 1987 figures for all classes of consumer in 'heat supplied basis' and in original units were compared (ie Table 8 and Table 9 of the 1988 Digest of United Kingdom Energy Statistics were compared). The 'heat supplied basis' uses the therm as the basic unit. A therm ($=10^5$ British Thermal Units – BTU) is equivalent to 105.51 MJ. Hence, all consumptions in therms could be readily converted into PJ – the units used in this report.

For coal a consumption of 16.55 million tonnes or 4451 million therms was quoted. This implies a calorific value of 269 therms per tonne which, in metric units, becomes 28.38 MJ/kg. In fact, this was rounded down to a figure of 28.3 MJ/kg for the calculations of carbon dioxide emissions.

For oil a consumption of 53.70 million tonnes or 23 440 million therms was quoted. This implies a calorific value of 436 therms per tonne which, in metric units, becomes 46.0 MJ/kg. This is the calorific value assumed for the calculations of carbon dioxide emissions.

For gas a calorific value of 55.5 MJ/kg was used. This is a figure based on the assumption that natural gas is almost pure methane.

Carbon contents of and carbon dioxide emissions from fossil fuels

Exactly the same figures for carbon contents of fossil fuels were used for the United Kingdom as for the world. The carbon dioxide emission per PJ was also worked out in exactly the same way, except that the calorific values assumed were those discussed in the previous section. Hence, the carbon dioxide emission factors calculated for the United Kingdom were:

Coal: $(0.70 \times 44)/(28.3 \times 12)$ = 0.0907 million tonnes/PJ

Oil: $(0.87 \times 44)/(46.0 \times 12)$ = 0.0693 million tonnes/PJ

Gas: $(0.75 \times 44)/(55.5 \times 12)$ = 0.0495 million tonnes/PJ

For 1987 these factors were applied to primary energy consumptions as determined from delivered energy statistics and the primary fuel mix of Table 5. The figures were also applied directly to primary energy statistics (Table 3) for each year from 1950 to 1987. It will be noted that these factors differ slightly from those assumed for the world statistics.

Table 3 shows that in recent years the United Kingdom has imported some electricity. Imported electricity is conventionally treated as primary electricity and, in this report, in common with other primary electricity, it is assumed that it produces no carbon dioxide. This is, of course, not strictly true but since it derives from France (where electricity generation is about 70% nuclear) and it is such a small amount (less than 4% of electricity use), the approximation is quite reasonable. In addition, it could be argued that, since it is carbon dioxide emission by the United Kingdom which is being calculated, imports of electricity should not count.

(continued)

Assumptions concerning technically possible and cost-effective savings

The assumptions used regarding the quantification of technically possible and cost-effective improvements are summarised below for all the energy-saving measures considered in Part 2 of the text.

Energy-saving improvement	Technically possible	Cost-effective
Insulation improvements	All walls insulated	80% of all cavity walls insulated
	All lofts insulated to 150 mm	Lofts with <=25 mm insulated to 150 mm
	Full double glazing in all homes	Full double glazing in all homes
	Draught proofing in all homes	Draught proofing in homes with <80% rooms already treated
	80 mm insulation to all hot water tanks	All tanks with <50 mm insulated to 80 mm
Heating efficiency improvements	Condensing boilers in all gas centrally heated homes	Condensing boilers in all gas centrally heated homes
Improvements to cookers	25% efficiency improvements to gas and electric cookers	13% efficiency improvements to gas and electric cookers
Improvements to lights and appliances	20% efficiency improvements to dishwashers	10% efficiency improvements to dishwashers
	75% reduction to lighting consumption	38% reduction to lighting consumption
	25% efficiency improvements to refrigeration equipment	25% efficiency improvements to refrigeration equipment
	20% efficiency improvements to washing machines <u>and</u> <u>tumble-driers</u>	20% efficiency improvements to washing machines
	25% efficiency improvements to televisions	25% efficiency improvements to televisions

ACKNOWLEDGEMENTS

This report draws extensively upon the results of work supported by both the Department of the Environment and the Department of Energy.

REFERENCES

1   Bolin B, Döös B R, Jäger J and Warrick R A (Editors). The greenhouse effect, climatic change and ecosystems. Chichester, John Wiley & Sons, 1986.

2   United Nations. 1980 energy statistics yearbook. New York, United Nations, 1981.

3   United Nations. 1985 energy statistics yearbook. New York, United Nations, 1987.

4   Department of Energy. Digest of United Kingdom energy statistics. London, HMSO, 1986.

5   Department of Energy. Digest of United Kingdom energy statistics. London, HMSO, 1988.

6   Department of the Environment. Digest of environmental protection and water statistics, No 11. London, HMSO, 1988.

7   Anderson B R, Clark A J, Baldwin R and Milbank N O. BREDEM - BRE Domestic Energy Model: background, philosophy and description. Building Research Establishment Report. Garston, BRE, 1985.

8   Henderson G and Shorrock L D. Energy efficiency in the housing stock. Building Research Establishment Information Paper IP22/88. Garston, BRE, 1988.

9   Hardcastle R. The pattern of energy use in the UK - 1980. Department of Energy. London, HMSO, 1984.

10 Wozniak S J. Solar heating systems for the UK: design, installation, and economic aspects. Building Research Establishment Report. London, HMSO, 1979.

11 Department of Energy. A study of passive solar housing estate layout. Contractor Report ETSU S 1126. London, Department of Energy, 1988.

12 Pezzey J. An economic assessment of some energy conservation measures in housing and other buildings. Building Research Establishment Report. Garston, BRE, 1984.

13 Leach G, Lewis C, Romig F, van Buren A and Foley G. A low energy strategy for the United Kingdom. The International Institute for Environment and Development. Science Reviews. London, 1979.

14 Henderson G and Shorrock L D. Domestic energy fact file. Building Research Establishment Report. Garston, BRE, 1989.

15 Hunt D R G and Steele M R. Domestic temperature trends. The Heating and Ventilating Engineer, 1980, 54 (626) 5-15.

16 **Trim M J B.** The performance of gas-fired condensing boilers in family housing. <u>Building Research Establishment Information Paper</u> IP10/88. Garston, BRE, 1988.

17 **Curwell S R, Fox R C and March C G.** <u>Use of CFCs in buildings.</u> Fernsheer Ltd, December 1988.

Note that after the figures were calculated, they were rounded for presentation in the tables which follow. This means that, in some places, sums of the rounded numbers may not exactly equal the totals given, although these 'rounding errors' are small.

TABLES

Table 1 Primary energy consumption (PJ) by world region (electricity includes nominal thermal efficiency of 30%; per capita figures are in GJ)

Primary energy consumption each year

World region	Fuel type	1950	1955	1960	1965	1970	1971	1972	1973	1974	1975	1976	1977	1978	1979	1980	1981	1982	1983	1984	1985
Africa	Solid	761.7	1 012.1	1 143.6	1 393.4	1 566.6	1 668.1	1 710.9	1 712.1	1 740.0	1 801.2	1 883.4	1 896.3	1 947.3	2 161.1	2 185.0	2 381.2	2 577.5	2 724.4	2 997.1	3 095.0
	Liquid	325.4	470.2	670.1	965.3	1 245.1	1 438.3	1 530.7	1 642.9	1 746.7	1 858.5	1 970.5	2 140.8	2 278.0	2 294.8	2 387.0	2 561.1	2 735.1	2 903.2	3 007.5	3 020.5
	Gas	0.1	0.2	0.9	43.1	15.4	32.3	87.1	110.9	107.4	134.4	185.7	137.0	578.9	259.4	314.7	403.1	491.4	753.4	686.2	735.2
	Electric	17.4	39.8	90.8	165.5	298.0	316.1	362.0	383.0	431.6	450.0	512.3	571.7	590.0	679.6	704.3	666.8	629.3	616.4	606.7	639.5
	Total GJ	1 104.6	1 522.4	1 905.3	2 567.3	3 125.1	3 454.8	3 690.8	3 848.9	4 025.6	4 244.1	4 551.9	4 745.7	5 394.4	5 395.0	5 591.1	6 012.2	6 433.3	6 997.5	7 297.5	7 490.2
	capita	5.1	6.2	7.1	8.3	8.8	9.5	9.9	10.0	10.2	10.5	10.9	11.0	12.1	11.8	11.9	12.3	12.6	13.4	13.5	13.5
USA/Canada	Solid	15 077.8	12 224.4	10 753.7	12 723.9	13 910.8	12 852.8	13 810.7	13 671.2	13 616.8	14 179.7	14 914.8	14 990.8	14 495.8	16 520.2	17 029.4	16 785.5	16 541.7	17 219.3	18 204.2	18 841.1
	Liquid	13 096.0	17 165.7	19 796.3	23 586.6	30 063.5	31 111.2	33 505.7	34 984.7	33 781.2	32 970.6	35 370.3	36 813.3	37 270.4	36 004.7	32 895.8	32 114.2	31 322.5	30 685.7	31 883.9	31 666.9
	Gas	6 561.4	9 892.2	13 876.1	17 757.7	24 516.8	25 448.9	25 628.5	24 629.1	23 048.2	23 491.1	22 361.7	23 751.9	24 261.1	23 751.9	23 860.8	21 991.6	20 231.3	20 420.6	20 131.0	19 834.6
	Electric	1 848.6	2 313.4	3 068.7	3 820.0	5 173.8	5 683.1	6 215.9	6 824.9	7 734.0	8 314.0	8 512.3	8 695.7	9 933.5	9 824.4	9 860.8	10 295.1	10 729.3	11 370.4	11 984.8	12 514.6
	Total GJ	36 383.7	41 595.3	47 494.7	57 888.1	73 664.9	75 095.9	79 487.4	81 109.2	79 761.1	78 512.4	82 288.6	83 461.4	85 110.2	86 610.3	83 537.9	81 186.4	78 834.8	78 695.9	82 203.9	82 857.2
	capita	220.4	229.1	239.2	270.5	327.2	329.6	345.5	349.7	341.2	332.3	345.5	347.5	352.1	355.3	332.1	319.5	307.2	304.0	313.6	314.5
Rest of America	Solid	238.6	268.3	285.1	308.6	375.5	375.1	380.1	408.0	458.3	483.2	506.8	563.4	633.2	664.8	676.1	655.5	634.9	711.7	762.0	845.7
	Liquid	1 362.9	1 982.5	2 745.4	3 433.1	5 090.6	5 585.4	5 900.4	6 433.5	6 686.4	6 751.1	7 143.0	7 565.1	8 033.9	8 401.1	8 706.4	8 614.9	8 523.3	8 323.7	8 030.3	8 252.7
	Gas	99.2	212.0	421.7	843.3	1 207.3	1 086.5	1 345.8	1 503.7	1 560.8	1 567.2	1 593.2	1 667.0	1 974.0	2 287.0	2 445.4	2 496.5	2 547.7	2 601.8	2 653.5	2 738.6
	Electric	159.0	262.3	420.0	613.5	981.0	1 052.1	1 187.1	1 314.6	1 496.4	1 612.7	1 804.3	1 979.4	2 141.6	2 388.1	2 621.1	2 825.9	3 030.7	3 224.3	3 543.6	3 790.9
	Total GJ	1 859.7	2 726.2	3 872.2	5 198.5	7 654.4	8 251.4	8 813.3	9 659.9	10 202.1	10 414.2	11 047.2	11 774.9	12 782.6	13 741.0	14 448.9	14 592.7	14 736.5	14 861.5	14 989.4	15 627.9
	capita	11.5	14.8	18.2	21.5	27.5	28.9	30.1	32.1	33.0	32.8	33.8	35.1	37.1	38.8	39.5	39.2	38.9	38.4	37.9	38.6
Asia	Solid	2 855.4	4 457.1	12 176.2	13 051.9	13 848.4	13 860.7	14 746.0	15 440.7	15 569.5	16 059.1	16 360.7	17 796.7	18 991.2	19 852.5	20 162.9	21 295.7	22 428.5	23 803.5	26 045.2	28 113.6
	Liquid	548.5	1 511.9	2 954.0	5 601.3	11 717.2	13 400.1	14 649.3	16 645.6	16 569.5	17 205.6	18 779.8	19 934.6	20 791.7	20 811.4	20 792.4	20 801.9	20 811.4	21 808.9	22 263.2	22 236.7
	Gas	37.9	124.6	282.3	680.9	2 074.4	2 275.5	2 380.3	2 274.8	2 668.6	2 939.5	3 240.2	3 056.6	3 572.9	4 025.6	4 445.7	4 736.8	5 027.9	5 532.7	5 825.9	6 699.2
	Electric	525.7	738.9	1 127.6	1 680.9	2 074.4	2 188.3	2 039.5	2 188.3	2 369.3	2 345.8	2 291.6	3 056.6	3 572.9	4 189.6	4 445.7	4 736.8	5 027.9	5 532.7	4 387.6	4 664.6
	Total GJ	3 967.4	6 832.5	16 540.6	18 454.0	27 930.0	30 656.0	32 157.0	35 218.8	36 525.3	38 245.3	40 569.0	43 157.2	46 039.6	48 189.9	48 868.7	50 202.7	51 536.7	54 790.9	58 521.8	61 714.2
	capita	2.9	4.6	10.1	10.2	13.5	14.4	14.9	16.0	16.2	16.6	17.3	18.1	19.0	19.5	19.4	19.3	19.1	20.0	21.0	21.8
Eastern Europe/ USSR	Solid	10 354.9	14 931.0	18 235.0	20 582.1	22 351.2	22 724.1	23 175.4	23 318.2	23 402.5	24 186.3	24 838.6	25 303.2	25 595.6	25 862.7	25 707.2	25 055.1	24 403.0	24 438.2	24 628.3	25 390.8
	Liquid	1 581.0	2 873.2	4 893.1	5 899.7	11 384.7	12 191.9	12 832.8	14 251.8	15 125.1	16 157.9	16 761.2	17 190.2	17 407.3	18 194.9	18 660.1	18 630.3	18 445.2	18 352.1	18 183.8	18 109.9
	Gas	381.7	625.5	2 329.9	8 099.9	8 048.3	9 032.3	9 492.0	10 106.8	10 773.2	11 801.6	13 033.5	13 959.8	14 672.9	15 551.6	16 353.7	17 399.5	18 660.6	19 655.4	21 396.2	22 967.0
	Electric	181.6	330.7	689.3	1 097.4	1 685.0	1 740.3	1 804.9	1 843.1	2 071.8	2 039.5	2 291.6	2 581.1	2 935.7	3 131.0	3 430.8	3 539.3	3 647.8	4 042.7	4 726.4	5 260.7
	Total GJ	12 499.2	18 760.4	26 147.3	35 679.1	43 469.0	45 688.5	47 305.1	49 519.9	51 372.5	54 185.3	56 924.8	59 034.3	60 611.6	62 739.2	64 092.4	64 624.3	65 156.2	66 488.4	68 934.8	71 728.4
	capita	46.5	64.8	83.9	107.2	125.0	130.2	133.6	138.8	142.7	149.2	155.5	159.9	162.8	167.3	169.7	169.3	169.3	171.4	176.2	181.4
Western Europe	Solid	13 327.4	15 471.1	14 759.0	14 180.5	13 172.0	11 389.0	10 413.5	10 713.2	10 484.6	9 956.1	10 653.4	10 051.9	10 220.2	11 085.9	11 307.5	11 149.1	10 990.7	11 039.7	10 583.0	11 357.3
	Liquid	1 893.4	3 743.9	6 888.0	13 285.7	21 441.1	22 276.4	22 832.8	25 272.4	23 747.8	22 636.7	24 268.6	23 909.3	24 286.8	25 813.9	23 933.5	22 873.9	21 814.3	21 322.2	21 533.4	21 091.2
	Gas	49.7	207.3	467.9	791.5	2 963.8	3 839.1	4 913.1	5 636.9	6 432.8	6 827.4	7 358.8	7 372.5	7 770.4	7 242.2	8 123.7	8 035.5	7 947.3	8 256.2	8 456.3	8 764.7
	Electric	1 328.3	1 912.6	2 791.2	3 776.9	4 715.7	5 089.8	5 232.4	5 676.2	5 819.3	6 038.5	6 674.7	7 023.3	7 216.4	7 782.2	8 410.8	9 039.3	9 955.9	10 300.7	11 300.7	12 376.0
	Total GJ	16 598.9	21 332.9	24 906.1	32 034.5	41 195.2	42 220.3	44 232.7	46 854.9	46 341.4	45 458.8	48 100.1	48 357.0	49 493.7	51 147.0	51 307.5? 51 147.0	50 469.3	49 791.6	50 574.1	51 873.5	53 589.2
	capita	54.9	68.1	76.3	93.5	116.2	118.3	123.1	129.6	127.5	124.4	131.2	131.4	134.1	142.4	137.4	135.4	133.3	135.2	138.4	142.9
Oceania	Solid	538.9	614.3	700.2	815.2	877.6	871.2	904.6	932.4	995.4	1 010.7	1 126.0	1 146.6	1 115.1	1 150.6	1 144.1	1 191.7	1 239.1	1 221.0	1 262.8	1 455.9
	Liquid	181.6	317.0	463.1	778.4	1 064.5	1 124.2	1 146.3	1 159.9	1 284.6	1 285.7	1 285.4	1 328.8	1 292.0	1 338.9	1 364.3	1 353.1	1 342.0	1 261.6	1 282.2	1 308.5
	Gas	0.0	0.0	0.0	0.1	60.1	88.7	127.4	142.1	355.6	182.4	222.5	280.2	392.8	324.4	370.9	453.7	536.6	548.5	602.6	579.2
	Electric	53.3	75.6	124.9	210.1	264.2	318.9	332.3	333.2	355.6	407.4	389.0	361.4	306.6	403.3	425.7	420.0	414.2	412.0	432.8	448.2
	Total GJ	773.9	1 010.9	1 288.3	1 803.8	2 266.4	2 403.0	2 510.5	2 567.6	2 804.0	2 886.2	3 022.9	3 117.0	3 106.5	3 217.2	3 305.0	3 418.5	3 532.1	3 443.1	3 580.4	3 791.8
	capita	61.6	77.7	81.8	103.1	117.2	120.9	123.9	125.0	133.9	135.4	140.0	142.5	140.1	143.3	145.4	147.4	149.5	143.5	147.5	153.9
World	Solid	43 154.7	48 974.9	58 053.3	60 663.6	64 305.5	63 728.6	64 255.9	65 501.2	66 138.3	67 676.3	70 283.6	71 749.0	72 998.6	77 297.9	78 212.3	78 513.9	78 815.6	81 157.8	84 482.5	89 099.3
	Liquid	18 988.8	28 066.0	38 410.0	55 750.2	82 006.7	87 127.5	93 381.6	100 390.9	98 941.3	98 866.1	105 578.9	108 882.1	111 360.1	113 305.3	108 680.1	106 949.4	105 218.6	104 657.6	106 184.5	105 686.5
	Gas	7 129.9	11 061.0	17 378.8	25 847.2	37 898.2	40 812.1	43 187.2	44 681.3	45 518.2	45 600.6	48 073.0	48 747.3	50 391.9	54 035.3	54 828.0	54 148.2	53 468.3	54 881.7	58 313.4	60 283.9
	Electric	4 114.0	5 675.9	8 312.4	11 364.2	15 094.6	16 101.8	17 372.2	18 205.9	20 434.1	21 803.2	22 568.9	24 269.1	26 782.9	28 125.6	29 270.7	30 894.7	32 518.6	35 154.4	38 420.8	41 729.2
	Total GJ	73 387.5	93 785.5	122 154.5	153 625.2	199 305.0	207 769.9	218 196.9	228 779.2	231 031.9	233 946.2	246 504.4	253 647.6	262 538.6	272 709.3	270 991.1	270 506.1	270 021.2	275 851.5	287 401.2	296 799.0
	capita	29.5	34.7	41.2	47.0	54.5	55.7	57.5	59.2	58.7	58.3	60.4	61.2	62.2	63.5	62.0	60.2	58.4	58.7	60.2	61.2

Table 2 Carbon dioxide emissions (in millions of tonnes) by world region (per capita figures are in tonnes)

World region	Fuel type	Carbon dioxide emissions each year																			
		1950	1955	1960	1965	1970	1971	1972	1973	1974	1975	1976	1977	1978	1979	1980	1981	1982	1983	1984	1985
Africa	Solid	67	89	100	122	137	146	150	150	152	158	165	166	171	189	191	209	226	239	263	271
	Liquid	25	36	51	73	95	109	116	125	133	141	150	163	173	175	182	195	208	221	229	230
	Gas	0	0	0	2	1	2	4	6	5	7	9	7	29	13	16	20	25	38	35	37
	Total CO_2	91	124	151	198	233	257	271	281	291	306	324	336	373	377	389	424	459	498	526	538
	CO_2/capita	0.42	0.51	0.56	0.64	0.65	0.70	0.72	0.73	0.74	0.75	0.78	0.78	0.84	0.83	0.83	0.87	0.90	0.95	0.98	0.97
	CO_2/PJ	0.083	0.082	0.079	0.077	0.074	0.074	0.073	0.073	0.072	0.072	0.071	0.071	0.069	0.070	0.070	0.070	0.071	0.071	0.072	0.072
USA/ Canada	Solid	1 321	1 071	942	1 115	1 219	1 126	1 210	1 198	1 193	1 242	1 307	1 313	1 270	1 447	1 492	1 470	1 449	1 508	1 595	1 650
	Liquid	997	1 306	1 506	1 795	2 288	2 368	2 550	2 662	2 571	2 509	2 692	2 801	2 836	2 740	2 503	2 444	2 384	2 335	2 426	2 410
	Gas	331	500	701	897	1 238	1 285	1 311	1 294	1 244	1 164	1 186	1 160	1 182	1 225	1 199	1 111	1 022	981	1 017	1 002
	Total CO_2	2 649	2 877	3 149	3 806	4 745	4 779	5 070	5 154	5 007	4 915	5 185	5 274	5 288	5 412	5 195	5 025	4 855	4 824	5 038	5 062
	CO_2/capita	15.95	15.83	15.86	17.79	21.07	20.98	22.04	22.22	21.42	20.80	21.77	21.96	21.88	22.20	20.65	19.78	18.92	18.64	19.22	19.21
	CO_2/PJ	0.072	0.069	0.066	0.066	0.064	0.064	0.064	0.064	0.063	0.063	0.063	0.063	0.062	0.062	0.062	0.062	0.062	0.061	0.061	0.061
Rest of America	Solid	21	24	25	27	33	33	33	36	40	42	44	49	55	58	59	57	56	62	67	74
	Liquid	104	151	209	261	387	425	449	490	509	514	544	576	611	639	663	656	649	633	611	628
	Gas	5	11	21	43	61	63	68	76	79	79	80	84	100	115	123	126	129	131	134	138
	Total CO_2	130	185	255	331	481	520	550	601	628	635	668	709	767	813	845	839	833	827	812	840
	CO_2/capita	0.80	1.00	1.20	1.37	1.73	1.82	1.88	2.00	2.03	2.00	2.05	2.11	2.22	2.30	2.31	2.26	2.20	2.14	2.05	2.07
	CO_2/PJ	0.070	0.068	0.066	0.064	0.063	0.063	0.062	0.062	0.062	0.061	0.061	0.060	0.060	0.059	0.059	0.058	0.057	0.056	0.054	0.054
Asia	Solid	250	390	1 067	934	1 143	1 213	1 214	1 292	1 353	1 407	1 433	1 559	1 664	1 739	1 766	1 866	1 965	2 085	2 282	2 463
	Liquid	42	115	225	426	892	1 020	1 115	1 267	1 261	1 309	1 429	1 517	1 582	1 618	1 582	1 583	1 584	1 660	1 694	1 692
	Gas	2	6	14	26	55	57	64	78	93	103	111	120	136	154	175	170	165	184	222	236
	Total CO_2	294	512	1 306	1 386	2 090	2 290	2 393	2 637	2 707	2 819	2 973	3 196	3 381	3 511	3 524	3 619	3 714	3 929	4 197	4 391
	CO_2/capita	0.22	0.35	0.80	0.77	1.01	1.08	1.11	1.19	1.20	1.23	1.27	1.34	1.39	1.42	1.40	1.39	1.38	1.43	1.51	1.55
	CO_2/PJ	0.074	0.075	0.079	0.075	0.075	0.075	0.074	0.075	0.074	0.074	0.073	0.074	0.073	0.073	0.072	0.072	0.072	0.072	0.072	0.071
Eastern Europe/ USSR	Solid	907	1 308	1 597	1 803	1 958	1 991	2 030	2 043	2 050	2 119	2 176	2 217	2 242	2 266	2 252	2 195	2 138	2 141	2 157	2 224
	Liquid	120	219	372	616	866	928	977	1 085	1 151	1 230	1 276	1 308	1 325	1 385	1 416	1 418	1 420	1 397	1 384	1 378
	Gas	19	32	118	298	406	456	479	510	544	596	658	705	741	785	826	879	931	993	1 081	1 160
	Total CO_2	1 047	1 558	2 087	2 717	3 231	3 375	3 486	3 638	3 745	3 944	4 110	4 230	4 308	4 436	4 493	4 491	4 489	4 530	4 622	4 762
	CO_2/capita	3.90	5.39	6.70	8.16	9.29	9.62	9.85	10.20	10.40	10.86	11.23	11.45	11.57	11.83	11.89	11.78	11.67	11.68	11.81	12.04
	CO_2/PJ	0.084	0.083	0.080	0.076	0.074	0.074	0.074	0.073	0.073	0.073	0.072	0.072	0.071	0.071	0.070	0.069	0.069	0.068	0.067	0.066
Western Europe	Solid	1 167	1 355	1 293	1 242	1 066	998	912	938	918	872	933	881	895	971	991	977	963	967	927	995
	Liquid	144	285	524	1 011	1 632	1 695	1 812	1 923	1 807	1 723	1 847	1 819	1 848	1 964	1 821	1 741	1 660	1 623	1 639	1 605
	Gas	3	10	24	40	150	194	248	285	325	345	372	372	392	416	410	406	401	417	427	443
	Total CO_2	1 314	1 651	1 841	2 293	2 848	2 887	2 973	3 146	3 051	2 940	3 152	3 072	3 136	3 352	3 222	3 123	3 024	3 007	2 993	3 043
	CO_2/capita	4.35	5.27	5.64	6.69	8.04	8.09	8.27	8.70	8.39	8.04	8.60	8.35	8.50	9.04	8.66	8.38	8.10	8.03	7.99	8.11
	CO_2/PJ	0.079	0.077	0.074	0.072	0.069	0.068	0.067	0.067	0.066	0.065	0.066	0.063	0.063	0.063	0.063	0.062	0.069	0.068	0.058	0.057
Oceania	Solid	47	54	61	71	77	76	79	82	87	89	99	100	98	101	100	104	109	107	111	128
	Liquid	14	24	35	59	81	86	87	88	98	98	98	101	98	102	104	103	102	96	98	100
	Gas	0	0	0	0	3	4	6	7	9	9	11	14	15	16	19	23	27	28	30	29
	Total CO_2	61	78	97	131	161	166	173	177	193	196	208	216	211	219	223	230	238	231	239	256
	CO_2/capita	4.86	5.53	6.13	7.47	8.32	8.37	8.53	8.62	9.24	9.18	9.62	9.86	9.54	9.76	9.80	9.93	10.06	9.61	9.83	10.40
	CO_2/PJ	0.079	0.077	0.075	0.072	0.071	0.069	0.069	0.069	0.069	0.068	0.069	0.069	0.068	0.068	0.067	0.067	0.067	0.067	0.067	0.068
World	Solid	3 780	4 291	5 085	5 314	5 633	5 583	5 629	5 738	5 794	5 928	6 157	6 285	6 395	6 771	6 851	6 878	6 904	7 109	7 401	7 805
	Liquid	1 445	2 136	2 923	4 243	6 241	6 630	7 106	7 640	7 529	7 524	8 035	8 286	8 475	8 623	8 271	8 139	8 007	7 964	8 081	8 043
	Gas	360	559	878	1 305	1 914	2 061	2 181	2 256	2 299	2 303	2 428	2 462	2 596	2 726	2 769	2 734	2 700	2 772	2 945	3 044
	Total CO_2	5 585	6 985	8 886	10 862	13 788	14 274	14 916	15 634	15 622	15 755	16 619	17 033	17 465	18 120	17 891	17 751	17 612	17 845	18 426	18 892
	CO_2/capita	2.25	2.58	2.99	3.32	3.77	3.83	3.93	4.05	3.97	3.93	4.07	4.11	4.14	4.22	4.09	3.95	3.81	3.80	3.86	3.89
	CO_2/PJ	0.076	0.074	0.073	0.071	0.069	0.069	0.068	0.068	0.068	0.067	0.067	0.067	0.067	0.066	0.066	0.066	0.065	0.065	0.064	0.064

Table 3 United Kingdom primary energy consumption (PJ) (per capita figures are in GJ)

| | Primary energy consumption | | | | | | | | |
| | Fossil fuels | | | Electricity | | | | | |
Year	Coal	Oil	Natural gas	Nuclear	Hydro	Imports	Total	Overall total	GJ per capita
1950	5259	672	0	0	22	0	22	5952	118.8
1951	5393	757	0	0	24	0	24	6174	122.8
1952	5338	789	0	0	24	0	24	6152	121.9
1953	5377	856	0	0	24	0	24	6258	123.6
1954	5555	950	0	0	32	0	32	6537	128.6
1955	5583	1056	0	0	24	0	24	6663	130.6
1956	5596	1147	0	0	32	0	32	6774	132.4
1957	5429	1120	0	5	36	0	41	6590	127.9
1958	5194	1440	3	2	36	0	39	6676	128.8
1959	4880	1710	3	12	36	0	49	6641	127.3
1960	5112	1997	3	22	41	0	63	7175	136.7
1961	4968	2153	3	27	51	0	78	7201	136.4
1962	4994	2370	3	36	51	0	87	7453	140.2
1963	5068	2578	5	61	44	0	104	7756	144.9
1964	4880	2819	11	80	46	0	126	7836	145.4
1965	4826	3115	34	148	53	0	201	8177	150.7
1966	4551	3367	32	192	53	0	245	8195	150.0
1967	4268	3596	55	218	66	0	284	8203	149.4
1968	4306	3795	127	248	46	0	294	8522	154.5
1969	4224	4094	248	260	39	0	299	8865	160.0
1970	4039	4400	472	231	56	0	286	9197	165.2
1971	3586	4435	760	240	44	0	284	9064	162.1
1972	3151	4757	1079	257	44	0	301	9288	165.6
1973	3423	4816	1166	245	49	0	294	9699	172.5
1974	3035	4464	1396	294	51	0	345	9239	164.3
1975	3089	4004	1461	265	49	0	313	8867	157.7
1976	3140	3936	1551	313	46	0	359	8987	159.9
1977	3158	4006	1657	347	49	0	396	9217	164.0
1978	3086	4086	1717	325	51	0	376	9265	164.9
1979	3336	4077	1876	335	53	0	388	9677	172.1
1980	3109	3561	1876	325	49	0	374	8919	158.3
1981	3042	3253	1902	332	56	0	388	8585	152.4
1982	2849	3259	1891	388	58	0	447	8446	150.0
1983	2870	3112	1973	439	58	0	498	8453	150.0
1984	2033	3965	2018	473	51	0	524	8541	151.3
1985	2710	3373	2171	536	51	0	587	8842	156.2
1986	2921	3303	2205	517	58	41	616	9046	159.4
1987	2991	3206	2266	481	51	114	646	9108	160.1

Table 4 United Kingdom carbon dioxide emission (in millions of tonnes, millions of tonnes per primary PJ, and tonnes per capita)

	Carbon dioxide emission					
	Primary fuel type					
Year	Coal	Oil	Natural gas	Total	$CO_2/$ PJ	$CO_2/$ capita
1950	477	47	0	524	0.088	10.4
1951	489	52	0	542	0.088	10.8
1952	484	55	0	539	0.088	10.7
1953	488	59	0	547	0.087	10.8
1954	504	66	0	570	0.087	11.2
1955	506	73	0	580	0.087	11.4
1956	508	80	0	587	0.087	11.5
1957	492	78	0	570	0.086	11.1
1958	471	100	0	571	0.086	11.0
1959	443	119	0	561	0.085	10.8
1960	464	139	0	602	0.084	11.5
1961	451	149	0	600	0.083	11.4
1962	453	164	0	617	0.083	11.6
1963	460	179	0	639	0.082	11.9
1964	443	195	1	639	0.081	11.8
1965	438	216	2	655	0.080	12.1
1966	413	233	2	648	0.079	11.9
1967	387	249	3	639	0.078	11.6
1968	391	263	6	660	0.077	12.0
1969	383	284	12	679	0.077	12.3
1970	366	305	23	695	0.076	12.5
1971	325	308	38	670	0.074	12.0
1972	286	330	53	669	0.072	11.9
1973	310	334	58	702	0.072	12.5
1974	275	310	69	654	0.071	11.6
1975	280	278	72	630	0.071	11.2
1976	285	273	77	635	0.071	11.3
1977	286	278	82	646	0.070	11.5
1978	280	283	85	648	0.070	11.5
1979	303	283	93	678	0.070	12.1
1980	282	247	93	622	0.070	11.0
1981	276	226	94	596	0.069	10.6
1982	258	226	94	578	0.068	10.3
1983	260	216	98	574	0.068	10.2
1984	184	275	100	559	0.065	9.9
1985	246	234	108	587	0.066	10.4
1986	265	229	109	603	0.067	10.6
1987	271	222	112	606	0.067	10.6

Table 6 Carbon dioxide emission associated with each unit of delivered energy in 1987

Fuel type	Million tonnes/PJ
Coal	0.092
Coke/breeze	0.105
Other solid	0.117
Natural gas	0.055
Electricity	0.231
Petroleum	0.084

Table 5 United Kingdom delivered energy consumption and primary energy equivalents (PJ) in 1987

Fuel type	Delivered energy	Primary energy equivalents							Primary ratio
		Coal	Oil	Gas	Nuclear	Hydro	Imports	Total	
Coal	469.6	477.1	0.0	0.0	0.0	0.0	0.0	477.1	1.02
Coke/breeze	223.8	258.8	0.0	0.0	0.0	0.0	0.0	258.8	1.16
Other solid	46.0	59.5	0.0	0.0	0.0	0.0	0.0	59.5	1.29
Coke oven gas	34.4	77.8	0.0	0.0	0.0	0.0	0.0	77.8	2.26
Natural gas	2045.1	0.0	2.3	2257.6	0.0	0.0	0.0	2259.9	1.11
Electricity	898.9	2119.4	219.5	7.9	481.5	51.0	115.2	2994.5	3.33
Petroleum	2472.9	0.0	2983.2	0.0	0.0	0.0	0.0	2983.2	1.21
Total	6190.7	2992.6	3205.0	2265.6	481.5	51.0	115.2	9110.9	1.47

Table 7 United Kingdom delivered energy consumption (PJ) by sector and by delivered fuel type in 1987

| | Delivered energy consumption (PJ) | | | | | | | | | |
| | Energy used by buildings | | | | | Other energy uses | | | | |
Fuel type	Industry	Public admin	Miscel-laneous	Domestic	Total buildings	Industry	Trans-port	Agri-culture	Total non-buildings	Total all uses
Coal	28.8	33.2	5.5	217.3	284.8	184.3	0.1	0.4	184.8	469.6
Coke/breeze	1.1	6.3	8.5	13.2	29.1	194.1	0.0	0.5	194.7	223.8
Other solid	0.3	0.0	1.5	42.6	44.4	1.6	0.0	0.0	1.6	46.0
Coke oven gas	3.2	0.0	0.0	0.0	3.2	31.2	0.0	0.0	31.2	34.4
Natural gas	133.4	139.9	172.6	1107.8	1553.8	488.2	0.0	3.1	491.3	2045.1
Electricity	44.6	66.4	169.0	335.7	615.8	257.3	11.1	14.8	283.2	898.9
Petroleum	102.1	113.7	56.6	103.6	376.0	282.2	1776.1	38.6	2096.9	2472.9
Total	313.5	359.6	413.7	1820.3	2907.0	1438.9	1787.3	57.4	3283.6	6190.7

Table 8 United Kingdom carbon dioxide emission (in millions of tonnes) by sector and by delivered fuel type in 1987

| | Carbon dioxide emission (million tonnes) | | | | | | | | | |
| | Buiding-related CO_2 emissions | | | | | Non-building CO_2 emissions | | | | |
Fuel type	Industry	Public admin	Miscel-laneous	Domestic	Total buildings	Industry	Trans-port	Agri-culture	Total non-buildings	Total all uses
Coal	2.6	3.1	0.5	20.0	26.2	17.0	0.0	0.0	17.0	43.3
Coke/breeze	0.1	0.7	0.9	1.4	3.1	20.4	0.0	0.1	20.4	23.5
Other solid	0.0	0.0	0.2	5.0	5.2	0.2	0.0	0.0	0.2	5.4
Coke oven gas	0.7	0.0	0.0	0.0	0.7	6.4	0.0	0.0	6.4	7.1
Natural gas	7.3	7.7	9.5	60.7	85.1	26.7	0.0	0.2	26.9	112.0
Electricity	10.3	15.3	39.1	77.6	142.4	59.5	2.6	3.4	65.5	207.8
Petroleum	8.5	9.5	4.7	8.7	31.5	23.6	148.6	3.2	175.4	206.9
Total	29.6	36.2	54.8	173.4	294.1	153.8	151.2	6.9	311.8	605.9

Table 9 Energy consumption of United Kingdom dwellings (PJ) by end use and by delivered fuel type in 1987

| | Energy consumption (PJ) | | | | |
| | End uses | | | | |
Fuel type	Space heating	Water heating	Cooking	Lights and appliances	Total
Solid fuel	183.1	87.4	2.7	0.0	273.2
Gas	797.4	226.9	83.5	0.0	1107.8
Electricity	60.5	57.0	34.2	184.1	335.7
Petroleum	77.7	24.8	1.2	0.0	103.6
Total	1118.7	396.1	121.5	184.1	1820.3

Table 10 Carbon dioxide emission (in millions of tonnes) of United Kingdom dwellings by end use and by delivered fuel type in 1987

Fuel type	Carbon dioxide emission (million tonnes)				
	End uses				
	Space heating	Water heating	Cooking	Lights and appliances	Total
Solid fuel	17.7	8.4	0.3	0.0	26.4
Gas	43.7	12.4	4.6	0.0	60.7
Electricity	14.0	13.2	7.9	42.6	77.6
Petroleum	6.5	2.1	0.1	0.0	8.7
Total	81.9	36.1	12.8	42.6	173.4

Table 11 Possible energy savings through the application of technically possible or cost-effective insulation and draught proofing measures to United Kingdom dwellings

Fuel type	Fuel savings (PJ)					
	Energy-saving measures					
	Loft insulation	Wall insulation*	Double glazing	Draught proofing	Water tank insulation	Total savings
(a) Technically possible measures						
Solid fuel	8.6	54.0	16.3	10.2	8.7	97.9
Gas	37.3	235.3	71.1	44.5	22.7	410.9
Electricity	2.8	17.9	5.4	3.4	5.7	35.1
Petroleum	3.6	22.9	6.9	4.3	2.5	40.3
Total	52.4	330.1	99.7	62.4	39.6	584.2
(b) Cost-effective measures						
Solid fuel	5.9	26.7	16.3	10.1	5.8	64.8
Gas	25.6	116.2	71.1	44.2	15.1	272.1
Electricity	1.9	8.8	5.4	3.4	3.8	23.3
Petroleum	2.5	11.3	6.9	4.3	1.7	26.7
Total	35.9	163.0	99.7	62.0	26.4	386.9

*Technically possible measures include insulation of *all* wall types, whereas cost-effective measures include only those cavity walls that would be suitable for cavity wall insulation.

Table 12 Potential reductions in carbon dioxide emission through the application of technically possible or cost-effective insulation and draught proofing measures to United Kingdom dwellings

	Reduction in CO_2 emission (million tonnes)					
	Energy-saving measures					
Fuel type	Loft insulation	Wall insulation	Double glazing	Draught proofing	Water tank insulation	Total savings
(a) Technically possible measures						
Solid fuel	0.8	5.2	1.6	1.0	0.8	9.5
Gas	2.0	12.9	3.9	2.4	1.2	22.5
Electricity	0.7	4.1	1.2	0.8	1.3	8.1
Petroleum	0.3	1.9	0.6	0.4	0.2	3.4
Total	3.8	24.2	7.3	4.6	3.6	43.5
(b) Cost-effective measures						
Solid fuel	0.6	2.6	1.6	1.0	0.6	6.3
Gas	1.4	6.4	3.9	2.4	0.8	14.9
Electricity	0.4	2.0	1.2	0.8	0.9	5.4
Petroleum	0.2	0.9	0.6	0.4	0.1	2.2
Total	2.6	11.9	7.3	4.5	2.4	28.8

Table 13 Possible energy savings and carbon dioxide emission reductions through replacement of conventional domestic gas boilers by condensing boilers, assuming that technically possible or cost-effective insulation measures are applied

End use	Potential energy savings (PJ)	CO_2 reduction (million tonnes)
(a) With technically possible insulation		
Space heating	37.0	2.0
Water heating	32.8	1.8
Total	69.8	3.8
(b) With cost-effective insulation		
Space heating	64.0	3.5
Water heating	35.1	1.9
Total	99.1	5.4

Table 14 Energy use in dwellings for lights and appliances and possible savings through technically possible or cost-effective efficiency improvements

Appliances	Typical appliance consumption (kWh/year)	Ownership level (%)	National energy consumption (PJ)	Technically possible improvements			Cost-effective improvements		
				Efficiency improvement (%)	Energy saving (PJ)	CO_2 reduction (million tonnes)	Efficiency improvement (%)	Energy saving (PJ)	CO_2 reduction (million tonnes)
Washing machines	200	86	14.2	20	2.8	0.7	20	2.8	0.7
Tumble-driers	300	31	7.6	20	1.5	0.4	0	—	—
Dishwashers	500	7	2.9	20	0.6	0.1	10	0.3	0.1
Refrigerators	300	57	14.4	25	3.6	0.8	25	3.6	0.8
Fridge/freezers	740	43	26.1	25	6.5	1.5	25	6.5	1.5
Freezers	740	39	23.4	25	5.9	1.4	25	5.9	1.4
Kettles	250	86	17.9	—	—	—	—	—	—
Irons	75	98	6.1	—	—	—	—	—	—
Vacuum cleaners	25	98	2.0	—	—	—	—	—	—
Televisions	235	98	19.2	25	4.8	1.1	25	4.8	1.1
Lighting	360	100	30.1	75	22.6	5.2	38	11.4	2.6
Miscellaneous	240	100	20.0	—	—	—	—	—	—
Total	—	—	184.1	—	48.3	11.2	—	35.3	8.2

Table 15 Cooking consumption in dwellings and possible savings through technically possible or cost-effective efficiency improvements

		Technically possible improvements			Cost-effective improvements		
Fuel type	Energy consumption (PJ)	Efficiency improvement (%)	Energy saving (PJ)	CO_2 reduction (million tonnes)	Efficiency improvement (%)	Energy saving (PJ)	CO_2 reduction (million tonnes)
Solid fuel	2.7	0	0.0	0.0	0	0.0	0.0
Gas	83.5	25	20.9	1.1	13	10.9	0.6
Electricity	34.2	25	8.5	2.0	13	4.4	1.0
Petroleum	1.2	0	0.0	0.0	0	0.0	0.0
Total	121.5	—	29.4	3.1	—	15.3	1.6

Table 16 Summary of potential savings through the application of technically possible or cost-effective energy efficiency measures in United Kingdom dwellings

Energy efficiency improvement applied	Energy saving (PJ)	CO_2 reduction (million tonnes)
(a) Technically possible measures		
Improved insulation standards	584.2	43.5
Improved heating efficiencies	69.8	3.8
Improvements to cookers	29.4	3.1
Improvements to lights and appliances	48.3	11.2
Total	731.7	61.6
(b) Cost-effective measures		
Improved insulation standards	386.9	28.8
Improved heating efficiencies	99.1	5.4
Improvements to cookers	15.3	1.6
Improvements to lights and appliances	35.3	8.2
Total	536.6	44.0

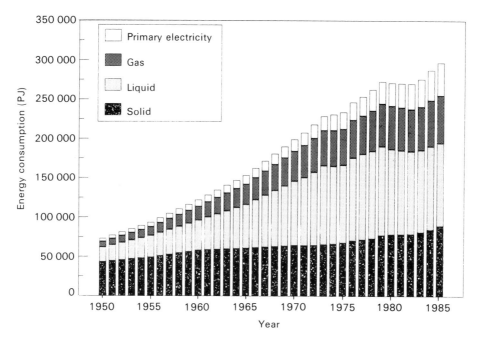

Figure 1 World primary energy consumption by fuel type

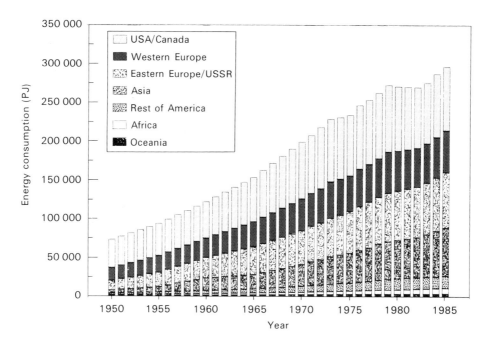

Figure 2 Primary energy consumption by world region

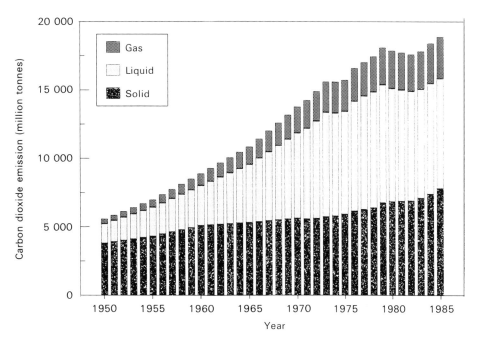

Figure 3 World carbon dioxide emission by primary fuel type

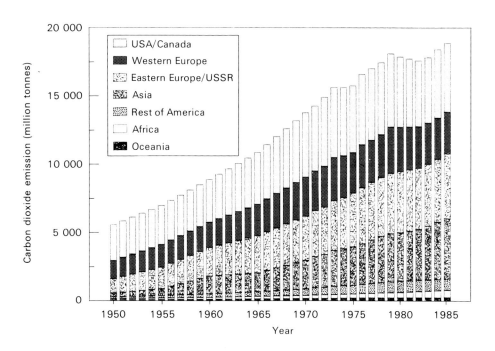

Figure 4 Carbon dioxide emission by world region

41

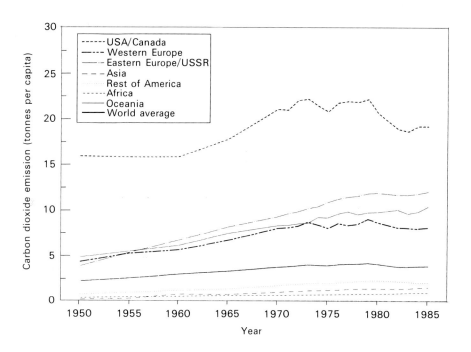

Figure 5 Carbon dioxide emission per capita by world region

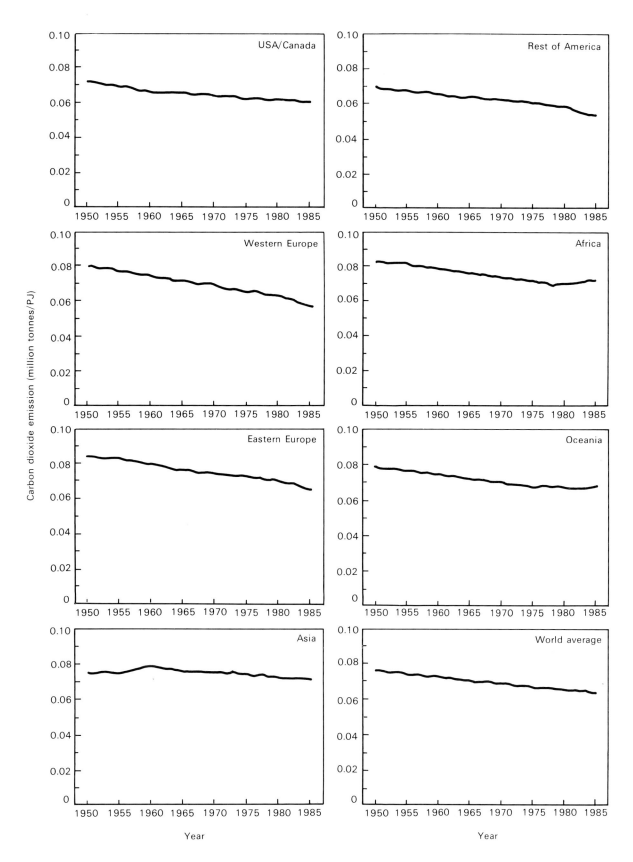

Figure 6 Carbon dioxide emission per primary PJ by world region

43

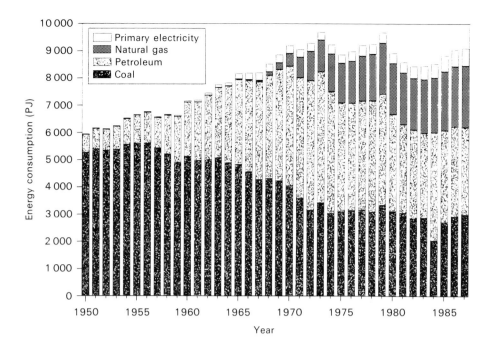

Figure 7 United Kingdom primary energy consumption

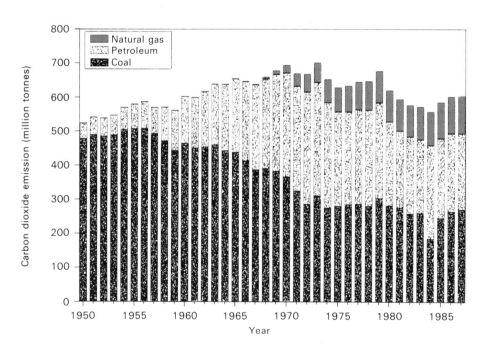

Figure 8 United Kingdom carbon dioxide emission by primary fuel type

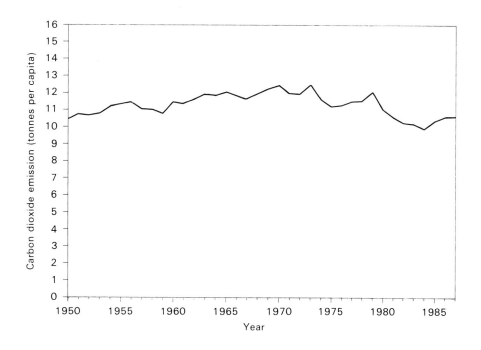

Figure 9 United Kingdom carbon dioxide emission per capita

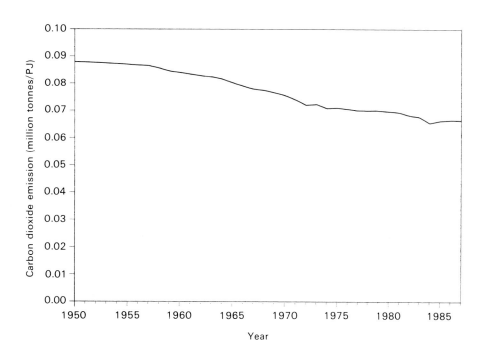

Figure 10 United Kingdom carbon dioxide emission per primary PJ

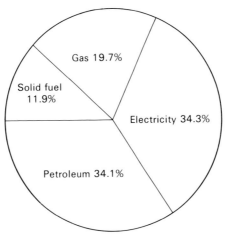

Total emission 606 million tonnes

Figure 11 United Kingdom carbon dioxide emission
by delivered fuel type in 1987

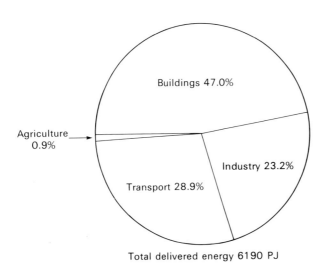

Total delivered energy 6190 PJ

Figure 12 United Kingdom delivered energy consumption by sector in 1987

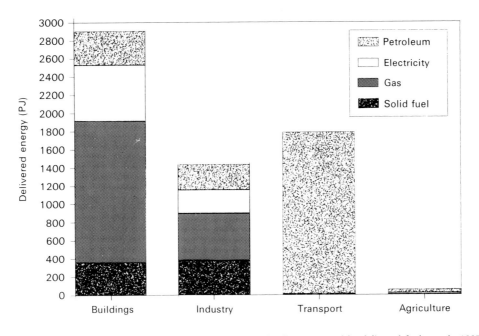

Figure 13 United Kingdom delivered energy consumption by sector and by delivered fuel type in 1987

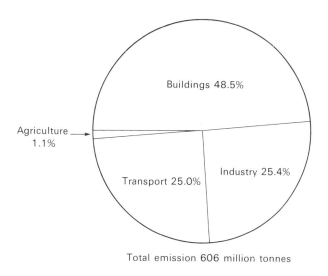

Total emission 606 million tonnes

Figure 14 United Kingdom carbon dioxide emission by sector in 1987

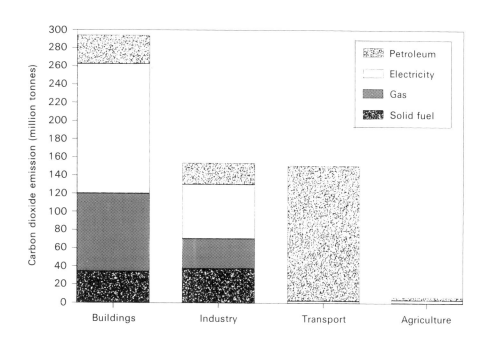

Figure 15 United Kingdom carbon dioxide emission by sector and by delivered fuel type in 1987

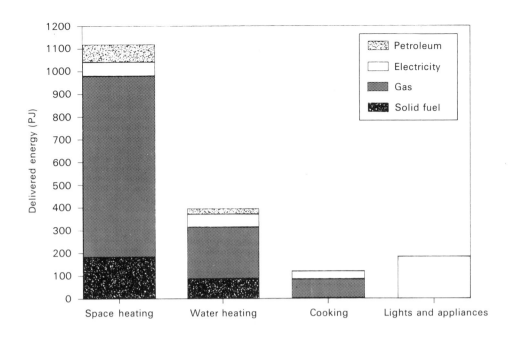

Figure 16 End uses of energy by delivered fuel type in United Kingdom dwellings in 1987

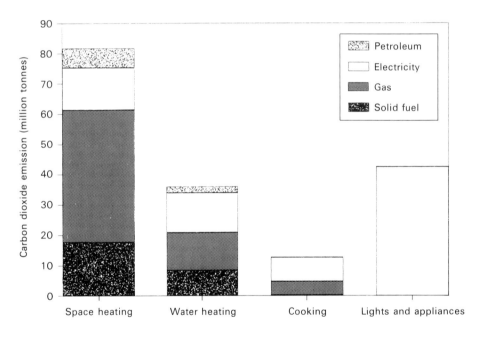

Figure 17 Carbon dioxide emission attributable to United Kingdom dwellings, by end use and by delivered fuel type in 1987

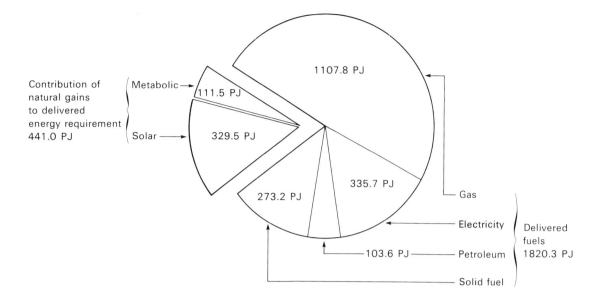

Total delivered energy 2261.3 PJ

Figure 18 United Kingdom housing stock delivered energy, including contribution by natural gains, in 1987

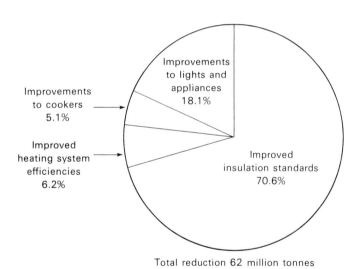

Total reduction 62 million tonnes

Figure 19 Carbon dioxide emission reductions through the application of technically possible energy efficiency measures in United Kingdom dwellings

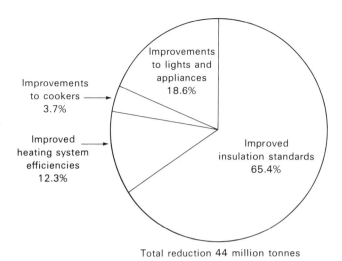

Total reduction 44 million tonnes

Figure 20 Carbon dioxide emission reductions through the application of cost-effective energy efficiency measures in United Kingdom dwellings

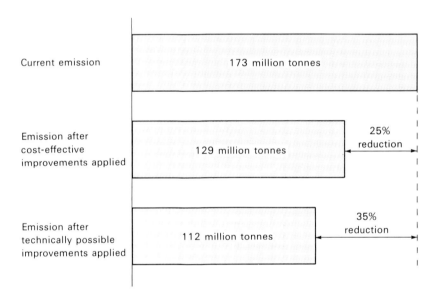

Figure 21 The impact of energy efficiency improvements on the carbon dioxide emission attributable to United Kingdom dwellings

50

Printed in the UK for HMSO. Dd.8245075, 3/90, C5, 38938.